808.023
M298t

Technical literature search and the written report

D J MALTHA

ELSEVIER
NEW YORK

First published 1972 by Pudoc, Wageningen
English edition first published 1976 by Pitman Publishing Ltd
Translated by J C Rigg

Published in the USA by:
AMERICAN ELSEVIER PUBLISHING COMPANY INC
52 Vanderbilt Avenue, New York, N.Y. 10017

© Pudoc, Wageningen 1976
ISBN 0 444 19501 7
LCCCN 76–21978

All rights reserved. No part of this publication
may be reproduced, stored in a retrieval system,
or transmitted, in any form or by any means,
electronic, mechanical, photocopying, recording
and/or otherwise, without the prior written
permission of the publishers.

Printed in Great Britain

Preface

This book is the result of a mental weakness: I cannot do two things at once. I was therefore in no position during my education both to listen to a lecture and to take down notes.

When as a lecturer at the Agricultural University, my turn came to give lectures, I assumed perhaps unjustly that such mental weakness was no exception amongst students. I therefore sought to establish the contents of each lecture in a syllabus, and to distribute copies of it in advance to the students. After some time a cupboard full of stencils had come into existence, out of which I could easily assemble various programmes for courses.

A few years later when the number of students that took the course suddenly grew to several hundred, it was essential to gather together the stencilled syllabuses and produce them in typed offset form solely for the use of students at the Agricultural University. A few colleagues who also received the lectures thought—as model publishers—that a wider distribution would be sensible. This has now happened and the lecture notes have been revised for general use by all who have to search the literature and write reports.

One difficulty remained, because many examples are from agricultural science: it has been impossible to replace them all by examples from other disciplines. It does not seem an insurmountable problem for users, and perhaps has the attraction of novelty to many non-agriculturalists.

The book is a guide, not a scientific treatise, aimed at all who, like me, struggle with paperwork.

The original book was written for students and other readers in the Netherlands. When the manuscript was translated by my colleague and friend Christopher Rigg, it became evident that many parts had to be revised and renewed for international use. This translation is an adapted version of the original.

Wageningen, August 1976   *D. J. Maltha*

Contents

Preface		v
1	**The information chain**	**1**
1.1	The information chain in the past	1
1.2	The information chain now	2
1.3	The future information chain	4
1.4	The chain of information to industry and commerce	5
	Further reading	6
2	**Purpose of the literature survey**	**7**
2.1	The publication as a medium in the information chain	7
2.2	Aim of literature searching	8
2.3	Documentary analysis	9
2.4	Literature searching	13
	Further reading	15
3	**Organizations for tracing literature**	**16**
3.1	Libraries	16
3.2	Documentation services	20
3.3	Information services	23
	Further reading	25
4	**Primary publications and bibliographical sources**	**26**
4.1	Primary publications	26
4.2	Bibliographical sources	28
4.3	Survey of primary and secondary publications	28
4.4	Current awareness service	49
	Further reading	54
	References	54
5	**Retrieval of literature**	**57**
5.1	General orientation	57
5.2	The systematic search	59
5.3	Rounding off the literature search	68

	5.4	Summary	69
		Further reading	70
		References	70
6	Recording information from the literature		71
	6.1	Descriptive and critical reports	71
	6.2	The card system	72
	6.3	Abstracting	77
		Further reading	82
7	Writing the literature survey		83
	7.1	Different procedures for writing	83
	7.2	Principles of text arrangement	86
	7.3	Some basic rules	87
	7.4	Subheadings and section numbering	90
	7.5	Use of language	91
	7.6	Factors in readability	94
	7.7	Readability research	98
	7.8	Use of illustrations and tables	99
	7.9	References to the literature	100
	7.10	Rounding off the report	103
	7.11	Some standards and other guidelines	105
	7.12	Worked example of the systematic method	111
		Further reading	114
8	The research report		115
	8.1	The scheme of research	115
	8.2	Scheme of the research report	118
	8.3	General rules	124
	8.4	Report of research on a method of analysis	125
	8.5	The scope: defining the problem	126
	8.6	The scientist's train of thought in the creative stage of research	134
	8.7	Expression of numeric data	139
		Further reading	157
		References	157
9	Disseminating the results of research		159
	9.1	Lines of communication	160
	9.2	Types of advisory publications and their characteristics	163
		Further reading	167
		References	167
		Index	169

1 The information chain

Every scientist forms a link in an information chain. On one side, he receives information on scientific advances in his field and other fields for which he has an interest. On the other side, he processes results of his own experiments into information for the benefit of other scientists.

This chain has existed for as long as there have been scientists. But in the past, it received little attention since the transfer of information raised no problems.

1.1 The information chain in the past

Forty years ago, scientists in a certain field would know one another, whether personally or from correspondence. A scientist considered it his duty to follow progress in his field by reading a few learned journals, and by exchanging reprints. He could keep all the information in his head; it was his spiritual baggage. Reference books completed what he could not manage by these channels. In material form, he had his spiritual baggage at home in his study, which was full of bookcases containing his private library. Only occasionally did he need to visit an academic or special library to study literature that he did not possess himself.

Since he knew the names of most of his colleagues, he could

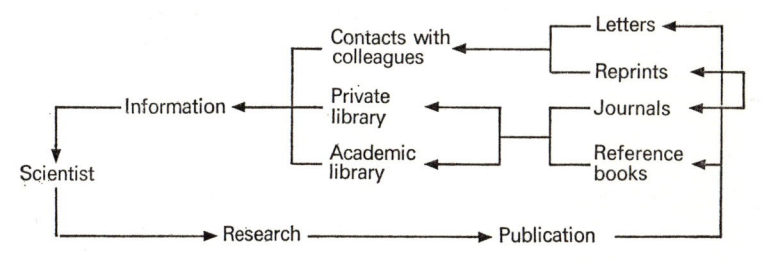

Fig. 1. Scheme of the information chain until about 1935.

merely consult the author catalogue in this library to trace work he wished to consult. There was hardly need for a systematic means of retrieving the limited literature. The information chain was not complicated and can be sketched as in Fig. 1.

1.2 The information chain now

Since then the situation has changed radically. There are various reasons for this change. The most important are as follows:
(*i*) The progress of scientific knowledge has been coupled with considerable specialization and at the same time coupling has occurred between branches of science. For the research scientist, it is no longer possible to see the whole picture.
(*ii*) The amount of literature has increased so much that it is impossible for anyone to keep up to date, even in a limited field, by reading a couple of journals and an occasional book.
(*iii*) It is not possible to survey the whole field of publishing. There are now tens of thousands of learned journals (figures mentioned are 40 000–60 000) and apart from these there are innumerable series, congress reports, annual reports and technical reports. Reference books have lost ground; instead, there are many monographs. The report literature has increased frighteningly. This literature is even less co-ordinated than journals.
(*iv*) Besides the classical Western scientific centres—publishing in Western languages—other centres have grown up publishing in languages inaccessible to the West. For these reasons the need came for a link between published material and the scientist: a record of what was published in such a systematic manner that any information required could be obtained at any moment: documentation. So came the need for specialists—documentalists—to establish these documentation systems.

The change shifted the problem of the accessibility of the literature to documentation and not surprisingly it is the documentalists who have dealt with the problem of making the literature accessible. This has lead to the construction of classifications, building up of card files, publishing of abstract journals, compiling of bibliographies, setting up of documentation centres, founding of national and international centres for translations and such like.

Another consequence has been that the scientist's private library has become relatively less important as a source of information and that he has been forced more and more to use the large academic libraries, or special institute libraries.

Initially the scientist considered that he could himself trace the literature he needed on new research through the documentation systems. But as the amount of literature increased, as specialization went further and as documentation systems became more complicated, a new category of specialists arose who assisted the scientist in tracing literature or took over the whole task. These specialists, known as information officers, must have a knowledge of their special field and must also be trained as follows:

to translate the scientist's question into a form suitable for tracing the literature in the documentation systems;

to handle the available documentation systems;

to select relevant information from the crude information obtained;

to convert this information into a form that the scientist can use. Fig. 2 is a scheme of how this system operates.

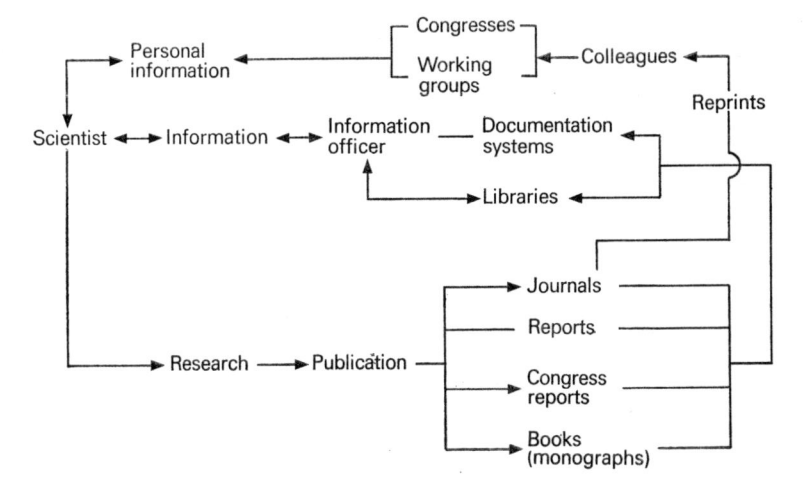

Fig. 2. Scheme of the information chain now.

Because of this change, the scientist no longer needs to know everything about all the systems that can provide him with information. But for good collaboration between the scientist and the information officer, it is essential that the scientist is aware of how documentation can help. During his scientific training, he must therefore receive a training in handling documentation systems and bibliographic sources, how to use libraries and how to process the information found.

1.3 The future information chain

With the advent of sorting and calculating equipment, people tried to mechanize a small part of the documentation work, in particular the sorting. It succeeded, but the machines were not fast enough.

Now that these sorting and calculating machines have been largely replaced by computers, people have tried to transfer a large part of the documentation and information work to these. We can say that they have succeeded in mechanizing all routine, repetitive or administrative operations. This is true, for instance, of the following operations:

  the systematic storage of information in the computer memory;
  the preparation of indexes from the stored information;
  the printing out of these indexes;
  the alerting of clients to information that is of probable interest to them, as judged by their profile (Selective Dissemination of Information, SDI).

The next phase is already in sight: the direct relation between the user of information and the information stored in the computer's memory, self-service from the computer. The future scheme will then look something like Fig. 3.

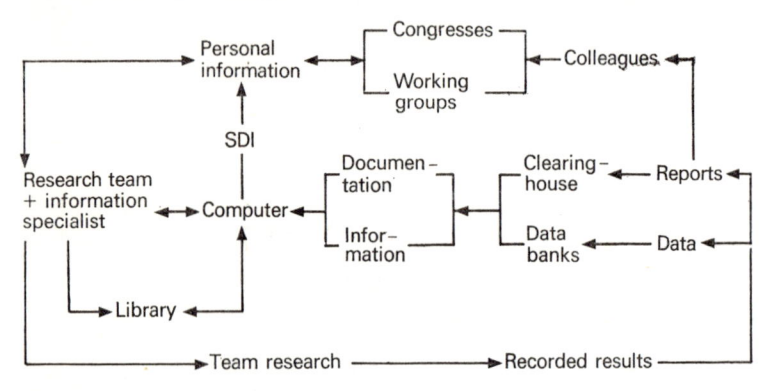

Fig. 3. Scheme of the future information chain.

The scheme does not include books and journals. This does not mean that we expect books and journals to disappear from scientific information, but they will be less important in the *direct* chain of information. Very probably they will hold ground as a *general* source of information: to keep the scientist aware of what is going on in a broad field. They will, so to speak, provide a counter-balance to overspecialization.

We can imagine that research will not be recorded primarily in journal articles, but in separates, which will be made available systematically by clearing-houses and computers. The same sort of function can be imagined for data banks, organizations where separate numerical data (for instance physical constants, structural formulae of chemical compounds) can be centrally stored and made available. This advanced system still has two weak points: the beginning or input; and the end or output.

Clearly, the usefulness of any documentation system as a source of information is dependent on how the information is taken from the original document. Unless the scientist agrees to record his results in a rigidly formalized scheme, direct machine processing of the document containing the results will give difficulties. Documentary analysis remains an operation in which people with their flexible mental abilities are irreplaceable. The furthest stage of that now operational is automatic analysis of the *titles* of documents. But this is often not enough.

Since mental work is still necessary for analysing and recording the information by abstracting or deep indexing, the document (for instance a research report) must meet high standards. In other words, the scientist-author is required to prepare a report meeting high standards. For the scientist himself, it is equally essential that his reports can stand the test of criticism. If he does not take enough care, the chance is great that his publication will be ignored as a source of information.

At the end of the line too, the human brain can rarely be missed. Machines can indeed be useful in collecting information, but experience shows that the specialist (scientist or information officer) must do the finer selection. Here as well, training is necessary.

The future scene is thus envisaged as a group of about 20 scientists with one information specialist who must ensure that scientists get the information.

1.4 The chain of information to industry and commerce

These schemes have so far considered only the flow of information in science from research to research. They have described the upward spiral of science. More is required of scientific research: that scientific results find application in industry or in the community. This is the purpose of development work or applied science.

The information chain must therefore contain an extension to practice. You could call this extension the vertical column of information. Information usually passes to practice in stages. In other words, scientific information does not go further than the management or the information service (as in agriculture). These groups ensure that the research results (adapted to commercial conditions) are applied. Fig. 4 gives this scheme.

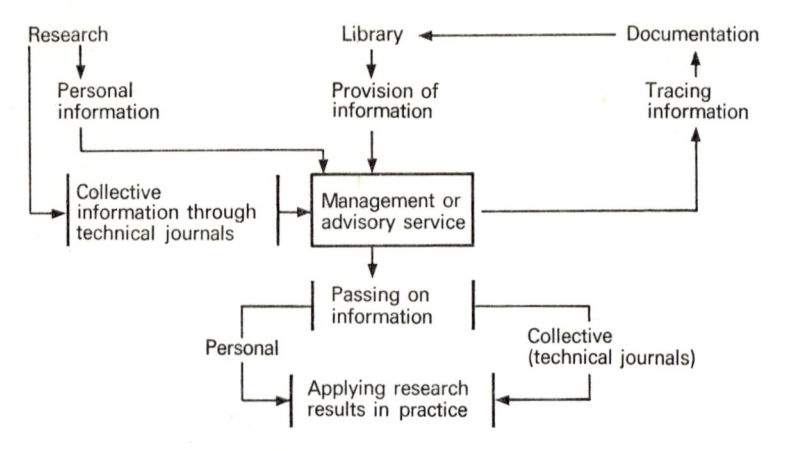

Fig. 4. Scheme of the information chain to industry and commerce.

In the following chapters we will not deal with all aspects of the information chain, but only with those which concern the scientist. In essence, there are three subjects:
   the tracing of literature on a given subject and the writing of a literature survey;
   the processing of research results into a research report;
   the passing on of research results to practice.

Further reading

Loosjes, Th. P. (1973). *On Documentation of Scientific Literature*, 2nd edn. Butterworths; London.
Vickery, B. C. (1973). *Information Systems*. Butterworths; London.

2 Purpose of the literature survey

The advancement of science means that people try to shift the frontier of knowledge. At this frontier one can see innumerable problems. The advancing of science amounts to choosing one problem from the many and attempting to solve it. Before beginning on this work one must be aware of how far knowledge of this problem has progressed, at least that knowledge which has already been published. This is always the starting point for one's own research.

The most important means of recording scientific information is still the publication. Publications take many forms; examples are: the reference book, the monograph, the journal article, the report, the congress report. All publications collectively are called the *scientific and technical literature*.

2.1 The publication as a medium in the information chain

A scientist must record his experimental results in a documented report of the research, in which he justifies the path he has followed to reach his results. His report must meet high standards of accuracy, completeness and logic. It is often preceded by a short preliminary report (preliminary note, short communication, letter to the editor) in which only results are given. The purpose of this is usually to gain priority (Section 4.3.1).

Reports take different forms but they are all *primary publications* (Section 4.1). Besides primary publications, there are secondary and tertiary publications. *Secondary publications* are derived from primary, and may take the form, for instance, of abstracts or review articles. There is an intimate relationship between documentation and information services and these secondary publications. Libraries and documentation and information services depend heavily on these secondary publications. You could even say that the literature would be completely inaccessible without this sort of publication.

Many secondary publications are made up and published by documentation services. Often documentation services and research collaborate closely in that the documentation service compiles literature and the scientist (specialist) writes abstracts or review articles.

Tertiary publications may be derived directly from primary publications or from secondary. They are characterized by describing practical applications of knowledge. If the writers of tertiary publications have set about their work properly, they will base their information on practical needs. This category includes innumerable trade journals.

2.2 Aim of literature searching

Information about a particular subject is usually traced by systematic study of the literature. This study of part of the literature is called *literature searching*. The purpose of literature searching is thus to collect information from the literature that gives a view of the state of the art at a certain moment within a certain defined subject.

Literature *study* covers a broader concept. It can have other purposes than literature searching. Examples are as follows:

(*i*) The study of literature on a new discipline in order to increase one's knowledge. For this one must concentrate on reference books, textbooks, monographs, review articles, reports, and proceedings of congresses.

(*ii*) Keeping up to date in a subject. For this one must concentrate on several current scientific or technical journals and must also be aware of the content of proceedings of congresses, review articles and abstract journals on that subject. Other important means of communication are exchange of reprints, exchange of letters between scientists and verbal discussion.

(*iii*) The broadening of one's general scientific knowledge. For this the reading of broadly oriented scientific journals, such as *Nature* (London), is important. Nowadays there are also scientific or popular scientific paperbacks, and even scientific news in newspapers such as *The Times*. The increasing interdependence between fields of science makes this sort of literature study essential.

We will confine ourselves further to literature searching. Before doing so, however, we must look at the aids available for retrieving literature.

2.3 Documentary analysis

It is no longer possible to search the literature without aids (Chapter 1). These aids can be termed (literature) documentation systems. The purpose of literature documentation is the collection, ordering and making available of literature data in a systematic manner so that information about the literature can easily be obtained from it.

Unfortunately the publishing world is not systematic. Especially among journals—quantitatively still the most important type of scientific publication—there are all possible combinations on national and international bases. De Solla Price (1965) calculated that the number of scientific journals has so far increased in a logarithmic curve. A journal is, in fact, a sort of linked sales package. The material is put together but not in the way that the scientist desires. Documentation's task is to take the journals apart (analysis) and to regroup the elements (the articles) bringing together the component parts for the scientific reader.

2.3.1 Analysis of collections of documents

Various organizations are concerned with the systematic collection of recorded knowledge. Academic libraries (e.g. university and Hochschule libraries) and special libraries (institutional libraries, commercial libraries) are important in this work. Particularly in industry, industrial archives with their collections of experimental reports form another important source of recorded knowledge. There are also the report archives (Chapter 1).

The duty of all these organizations is to make their collections accessible by cataloguing or by other means. In libraries, catalogues are the most important means. We will therefore limit ourselves to them. Most catalogues are in the form of card systems though there are also catalogues in the form of books or lists. Each card contains a description of a document (e.g. book, journal, brochure, report) present in the library and—usually in the form of a code—the place where the document can be found in the library. The description of the document is called the *bibliographical description* (catalogue entry; bibliographic reference; 'title'). A catalogue is thus a reflection on paper of the library's holdings and allows one to trace documents in these collections that are relevant without examining the collections of documents themselves.

Since it is becoming financially more difficult for each library to keep a collection reasonably complete on a certain subject, libraries are going over to sharing the responsibility for acquisition. This development has consequences for the cataloguing system. If a library decides not to buy, for instance, an expensive book because another library has agreed to buy it, a user in the first library would not be aware of its existence from that library's catalogue. Such collaboration thus leads to the compilation of central catalogues, joint catalogues or union catalogues listing the collections of several libraries.

Such collaboration can proceed at different levels. It is very common for the central library of a university to have a central catalogue for all the departments and institutions within the university. The next step is a national union catalogue of a particular subject, e.g. BUCOP (British Union Catalogue of Periodicals). The next step again is a general national catalogue. Particularly well known here is the National Union Catalogue of the Library of Congress in Washington. By modern techniques, for example, using the computer, the compiling and reproducing of such catalogues requires less work; and the exchange of such Union catalogues can even be arranged so that the sum of the material becomes, as it were, much greater than the amount of material the library itself contains. One could compare this development with giro systems for financial transactions.

Before using a library, one must find out what sort of catalogues it has. It can save a good deal of time and work spent on searching. Every library has an *alphabetic catalogue*. The alphabetic catalogue gives the bibliographic description arranged alphabetically according to a catchword. Usually the catchword is the name of the (first) author. But if there is no personal author or very many authors (for instance in collected works, a report of a commission or the name of a journal), the first noun or adjective of the title may be taken as the catchword.

If one has difficulty in looking up a title, there is always a librarian to ask. In large libraries, there is often an information desk. This is particularly necessary since libraries usually work with house rules that vary from one library to another.

The alphabetic catalogues are a good starting point if one is looking for a particular work and thus already knows: author's name, title of the book, name and address of the publisher and year of publication; or: name of a journal and volume that one wishes to consult. Remember that individual articles in journals and individual contributions to congress proceedings are not usually catalogued by libraries.

If one is looking for literature on a particular subject without knowing the name of any authors or the titles of publications, one must look up a *subject catalogue* (or *systematic catalogue*), or a *keyword catalogue*. Most libraries have one of these types as well as the alphabetic catalogue. To consult these one must know the system. It is usually described in a booklet or brochure or in a tabular scheme or list standing near to the catalogues. The systematic catalogue is a hierarchic classification of the discipline in which the library collects literature. This classification is sometimes expressed in words, sometimes as a code. The code can be made up of numerals or letters or combinations of numerals and letters.

A keyword catalogue contains words arranged alphabetically. Each word characterizes a subject. As a whole, the catalogue forms a list of technical terms for that discipline. How far it goes and how finely they are ultimately divided into headings (a secondary hierarchical or alphabetical order) is dependent on the size of the collection and on how much detail is desired in the keyword catalogue. For example: the keyword could be 'bibliographies and indexes' containing a second grade of keywords:

'in bibliographies';
'in dictionaries';
'in directories and yearbooks';
'in encyclopaedias';
'in periodicals'.

It is essential that both systematic and keyword catalogues remain usable by the ordinary reader. There should be enough cards, but never more than 50 behind each marker card. (See further Section 5.1 and 5.2)

Besides the common types of catalogue, there are special ones. Sometimes people use geographical or chronological catalogue entries or there are special catalogues for particular forms of publication (for instance proceedings of congresses, for periodicals in the library, for offprints or for commercial catalogues). There are even special catalogues for particular collections, for instance, of literature from particular countries or of early printed books.

A remarkable type of catalogue too is the *encyclopaedic* or *dictionary catalogue*. In this type, names of authors, chief words from the title and keywords of the keyword catalogue are arranged in one alphabetic system. This type of catalogue is common in the United States.

2.3.2 Analysis of works on a discipline

In using libraries and other collections, such as patents, documentation services attempt to analyse the recorded information on their subject. In their analysis, they usually go deeper than the library and use numerous aids such as card files, abstract journals, bibliographies, punched cards or tapes, magnetic cards or tapes and computer memories. This analysis has no purpose in itself but is a tool of information. Analysis must be directed to the provision of information. Retrieval from documentation systems works on the same principles as library catalogues. One can distinguish two groups: the group of retrieval systems based on hierarchic classification (the classification methods); and the group based on keywords (and an alphabetic arrangement of them).

Documentalists attempt to penetrate as deeply as possible into the subject matter of the publications. Their record is in very brief form (for instance as a code number or as one or more keywords) or in condensed form (for instance as abstracts) so that the user's retrieval of literature important for him can be as good and rapid as possible.

The principal difference between a library and documentation is that the library is based on a certain collection (books, journals etc.) and tries to make this collection accessible by bibliographic descriptions in catalogues, whereas documentation is based on the literature of a certain discipline and attempts to make it accessible by documentary analysis through documentation systems. There are many types of documentation systems which range from simple card systems to elaborate programs on computers, and also include abstract journals and other bibliographic publications. Library and documentation services are frequently combined. There is also commonly a functional and sometimes also an organizational division. Clearly whatever the organization, good co-ordination between the two types of service is necessary. The situation can be confusing for the reader especially if a library and a documentation service share the same building. He sees all sorts of card systems around him and can hardly detect which are library catalogues and which are card systems belonging to the documentation service. There is indeed only a qualitative difference: card systems in which each card contains more than just a bibliographic description, for example also an abstract, will usually belong to documentation. Such systems will usually contain more information than those of the library because they give access to a subject field. Analysis goes further too, because separate articles from journals, proceedings and other collected works are included in the system. However, separate housing of the two services is unattractive. The purpose of both services is to help the user to trace and obtain literature vital for him as adequately and as rapidly as possible.

2.4 Literature searching

The best place to start a literature search is in the documentation system rather than in the library catalogues, and the first thing to do is to look up the secondary publications. Reviews and abstracts can immediately provide important information. One can use card systems compiled by documentation services. One can seek information provided by mechanical means, for instance from magnetic tape, which can be searched by computer. One must certainly consider the costs of such work. Use of card systems and printed sources is generally free, but use of machine systems usually involve certain fixed costs. Remember too that no single system is complete in itself. One must simultaneously search in different directions.

With the information so collected one can approach the alphabetic catalogue of the library to try to obtain the original literature, whether as such or in reproduced form (for instance photocopy or microfiche). One of the library's functions is to help users to obtain material that the library does not itself possess. A common method is to send a request around various libraries; this system works slowly. Another way is to use telex instead of a request card. Most large libraries have telex connections. A third way is to use the existing central catalogues to find out who holds a certain work. Loans from libraries abroad are often still difficult: they can be channelled through the national library, e.g. British Library Lending Division.

In principle, one should check primary publications in compiling a literature review! If one is forced to use secondary publications (for instance because of unfamiliarity with the language of the primary publication), then one should mention the source in the literature survey. Obviously one should not use tertiary publications in compiling a literature survey. They are always derived and one is never sure whether the original information has been well reproduced. In literature searching one must therefore take serious account of the type of publication that one is examining. Fig. 5 gives the scheme of literature searching.

What the enquirer needs from his literature search is an overall awareness of it, so that he can be reasonably sure that he has a complete insight into the literature of his subject of interest. Usually it will be particularly important for him to know of recent advances in research; but he may also be interested in forming a picture of development over a longer period. Behind every literature search lurks the danger that the enquirer cannot stop searching. Unclarified questions remain; there are conflicting opinions, references to other

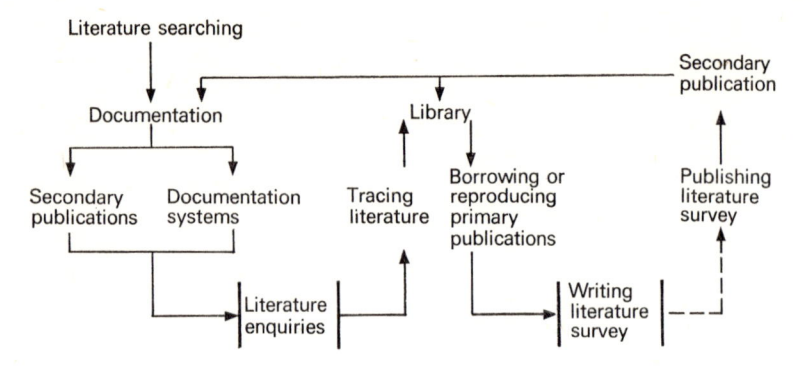

Fig. 5. Scheme of literature searching.

research, marginal subjects, etc. It is therefore essential to determine exactly, perhaps after a first approach to the literature:
 the purpose of the literature search;
 the subject on which literature is desired;
 the period for which literature must be collected.

The reason for the search can make quite a difference. Let us take some examples. If information is desired on the *progress* of research on a certain subject (for instance photosynthesis), it is not essential to know *all* the literature. One would certainly want to know the historical development and of course the latest developments. The important sources are encyclopaedias and reference books, proceedings of congresses, supplemented by recent literature from journals. If *new research* is being planned, it is not necessary to begin at the very beginning but certainly to check thoroughly whether that particular research or related research has been done earlier. One then defines the subject and goes thoroughly into the literature and tries to reach completeness within narrow limits. One may even try to find whether such research is in progress anywhere else (thus to check for current research).

If one is preparing a *thesis*, it is necessary to examine the question thoroughly by studying the literature and therefore surveying it critically. The literature survey of a thesis usually forms a separate chapter.

If one is preparing an *undergraduate thesis*, it is advisable at a first examination to keep the subject as narrow as possible and to examine the literature as deeply and thoroughly as possible. A good analysis of the literature is necessary for this; a critical survey of the literature is not always required but is certainly to be recommended.

If a subject that one is searching is only of *marginal interest*, a general orientation is sufficient. For instance, if one is preparing a paper on the effect of the national law of inheritance on the break-up of rural estates, one must examine the parallel literature in other countries. It is then sufficient to look at these only superficially.

When can the writer of a literature survey feel satisfied that he has an adequate view of the literature? There is no absolute criterion. An indication is when the literature listed in the most recent publications on the subject has already been found, and when the most recent numbers of journals that contain articles on the subject provide no new articles. It could well be that less accessible languages, such as Russian and Japanese, are still excluded. The only way of approaching them is to search the abstract journals or the bibliographies in which one would expect this literature to be processed (for instance East European abstract journals for the Russian literature).

Further reading

De Solla Price, D. J. (1965). *Little Science, Big Science*, 2nd edn. Columbia University Press; New York.

Peterson, M. S. (1961). *Scientific Thinking and Scientific Writing*. Reinhold Publishing Corp.; New York.

3 Organizations for tracing literature

Previous chapters have already mentioned the organizations whose purpose it is to make literature available: libraries and documentation services. These organizations are often co-ordinated or even combined into one organization. A sharp division is therefore impossible. Even so, there are differences in function and form of organization.

3.1 Libraries

The purpose of libraries is to supervise a collection of documents and to make them available to users. The 'library' is the organization, the building containing such a collection, or the collection itself. A library is thus a systematic collection of books, journals, patents, reports, separates, manuscripts, sheets of music or other written documents. Sound recordings (such as gramophone records or recorded tapes) can also form the collection of a library; likewise microfilm and microfiche.

Sometimes when the majority of the collection consists of media other than printed documents (for instance gramophone records, tape recordings, films and transparencies), the word media library is used instead.

The most important functions of a library are as follows:
acquisition: the growth of the collection (by purchase, exchange or gift); the management of the collection; preparation for storage, for instance by binding separate issues; the storage itself and the protection of the collection against damage;
cataloguing: specifications of the documents—bibliographical descriptions—and the construction of catalogues from them;
lending: making the documents available to people;
provision for study in reading rooms, carols or reading areas;
co-operation in obtaining documents or reproductions of documents from other libraries;

copying documents for those intending to use the copies for personal study.

In professional circles it is usual to divide libraries into three types:

learned libraries;
special libraries;
public libraries.

Some distinguish yet a fourth category: governmental libraries, such as the libraries of the ministries or the administration. This terminology is far from correct. Most learned libraries are open to the public; the distinction between learned libraries and special libraries is rather vague; many public libraries have scientific or technical branches; the majority of the research libraries are financed by the government. The division into these three types of libraries, however, can be used for practical reasons.

3.1.1 Research libraries

Most countries have a *national library*, sometimes with the name of royal library. Examples are: Österreichische Nationalbibliothek in Vienna, Bibliotheca Nazionale Centrale in Rome, Bibliothèque Royale in Paris, Koninklijke Bibliotheek in Brussels and in The Hague. The national library in the United Kingdom is the *British Library*, a department of the British museum in London; in USA it is the *Library of Congress* in Washington DC; both are very famous libraries.

The main tasks of these libraries are to collect and to maintain a collection of the *national* production of literature. Mostly this function is regulated by a law for legal depository (dépôt légal). The scope of the Library of Congress and the British Library, however, is much broader. The centralized national function is also expressed in the tasks of these libraries to control the bibliographical description of the national book production (national bibliography) and to execute the international loans and the international exchange service of publications. Sometimes national libraries have a mechanized service to give information on the location of books and journals in the other libraries of the country (central catalogues).

In the UK most of the national and international loans of publications are concentrated in the National Library Lending Division (NLL) at Boston Spa. NLL has a very large reproduction section sending photocopies and microfiche for study purposes all over the world.

Libraries of universities maintain large collections of nearly every discipline. They have the task of taking care of the coverage of the international scientific literature in the country. They have a function as a depository for this material. Moreover, it is their duty to promote the accessibility of these collections in such a way that easy retrieval for academic education and research is possible.

The faculties of a university usually have their own more specialized collection of their own discipline, but it is a rule that these collections belong to the total content of the university library. Many university libraries have a central catalogue or union catalogue. These contain not only the bibliographical descriptions of the titles of the publications in the main library, but also of the publications available in the faculty libraries or even of the specialized libraries of research institutes which do not belong to the university. The advantages for the user are evident: he has to look in one catalogue only to locate a desired publication.

The exponential growth of the scientific literature in the world poses the university libraries great problems. Their budgets are insufficient to maintain a satisfactory coverage of the international scientific literature. A collaboration between these libraries is now coming to the fore in the way that they split up their tasks. Each of the collaborating libraries accepts the responsibility for collection of the literature in one or more disciplines. Such collaboration has already existed for many years in the Scandinavian countries (the Scandia-plan) and the German academic libraries have worked out a similar plan (Schwerpunkte).

The largest university library in Western countries is the library of Harvard University (7 250 000 volumes), followed by that of Yale University (4 700 000 volumes). The University Library in Moscow has a collection of 6 million volumes. The largest and most important university libraries in Western Europe are those of Oxford, Cambridge, London, Paris, Strasbourg, Oslo, Upsala, Göttingen, Heidelberg and Basel. They all have collections of between 1.5 and 3 million volumes.

Another category of academic libraries is the group of libraries of learned societies, such as the academies of sciences. The function of these libraries is principally to serve the members, but some of these libraries also act as a public academic library.

3.1.2 Special libraries

The most characteristic feature of the special libraries is not that they have a specialized collection, but that they serve the needs of the staff of a firm, a laboratory, an organization, etc. The purpose of their services is to provide their clients with information. Especially in this category of libraries the functions of the library and of the documentation and information service are interrelated; they are combined in one unit.

In the selection of new books and journals for this collection the special librarian is guided by the actual needs of the organization to which the library belongs. He will try to update his collection continuously, but on the other hand he will dispose of obsolete material as soon as possible. In a manner of speaking his collection is advancing with time. The special library is not the right place to make historical studies.

This feature of special libraries means that the special librarian needs very close contact with his colleagues and with the large academic libraries, due to the fact that his own collection is restricted in the number of volumes and in the time period covered.

Many special libraries are organized on the principle of *free access*. The books on the shelves are arranged in a systematic order and the user can look for the literature of his interest without consulting the systematic catalogue. Free access is very useful for browsing, but on the other hand the disadvantage is that a book can be in one place on the shelves only, with the consequence that the user does not find the book on another shelf although it would be logical to find it on this shelf according to the systematic order of the library.

This defect is even more serious with collections of reprints that are made by many special libraries. One has to be cautious in consulting these collections. It is possible to collect very quickly a number of references to the subject as an orientation, but it is a rather arbitrary choice from the total amount of reference sources.

Special libraries collect, more than the other types of libraries, reprints, patents, priced catalogues, technical descriptions of apparatus and machinery, standards, annual reports, maps, time-tables, etc.

3.1.3 Public libraries

The function of public libraries is to provide the general public in their region with literature, and also with gramophone records and other audio-visual media. In this way the public libraries contribute to a justified spending of leisure time, but also the development and the education of large groups of people. The public library functions more and more as an information and educational centre in the region. Many public libraries have collections of scientific and technical books and journals. Their function is especially important for, what is called, 'éducation permanente'. They organize lectures, exhibitions, etc. according to their social function.

Public libraries originated in the United Kingdom (Norwich, 1850, Manchester, 1852) and in the United States (Boston, 1851). They are very developed in the Scandinavian countries and in Germany.

3.1.4 Governmental libraries

It is self-evident that the main task of the libraries of the administration is to serve the governmental agencies as a source of information, but especially some of the libraries of the ministries have a broader scope. A typical example is the library of the United States Department of Agriculture (USDA) which is the National Agricultural Library of the USA, the largest agricultural library in the world. The library of the Ministry of Agriculture, Fisheries and Food in London also has a central function.

3.2 Documentation services

By comparison with libraries, documentation services are 'young'. The oldest known library dates from about 2575 B.C. (the library of King Shepseskaf at Gîza in Egypt); the oldest known scientific journal with abstracts is *Pharmaceutisches Centralblatt* of 1830. It is therefore more difficult to generalize on the function of documentation services than of libraries. Their functions vary widely and they are still evolving. We can therefore only define some functions that occur in some but not all documentation services.

A common characteristic is that documentation services try to make the literature of a discipline or subject available. The services therefore pay special attention to separate articles in journals. Services usually provide a list of the journals they cover. The processing of other publications (reference books, textbooks, theses and reports) is often less accurately and completely defined.

The way in which these journals and other literature are processed is extremely diverse. At the one extreme, one could mention *Current Contents*, which exist for various subject fields. These secondary current publications (in the form of journals) are no more than a collection of the lists of contents from separate issues of a number of journals. They provide a rapid but superficial indication of the content of the issues of these journals.

The other extreme is the abstract journal. These secondary publications consist of abstracts of each relevant article in a discipline or subject prepared by a staff of experts. These abstracts give very briefly the essentials of the articles. Abstract journals are the best approach to the literature, but it must be remembered that abstract journals are selective (and do not include all the contents of the journals they scrutinize) and that abstracting delays the appearance of the information. The average time lag is nine months after the appearance of the original literature, and can vary from three months to several years.

Between these extremes there are various systems. There are current bibliographic publications containing only the titles of articles and books, retrievable by indexing to greater or lesser depth; there are publications with titles and keywords; there are publications with titles and author's summaries and those with titles and short annotations.

Besides these current publications there is another important category of descriptive bibliographic publication variously called 'reviews' or 'state-of-the-art surveys'. They are often published regularly (annually or once in two or more years); sometimes they are single publications.

For almost every subject field a list can be made of the most important secondary sources and it is therefore important that students attempt to get to know these sources during their study. It could be said that these sources have taken the place of the 'mental baggage', mentioned in Section 1.1.

Besides printed publications, documentation services supply magnetic tapes containing bibliographic information with or without abstracts. With such tapes and with the manual used to prepare them, it is possible to retrieve information by computer. The type of enquiry must, however, be adapted to what the system offers. In some cases, the documentation services themselves supply this information; otherwise the tapes may be processed by institutional, national or international computer centres. More and more documentation services are now supplying information by hire, lease or sale of magnetic tapes alongside printed publications. The provision of services that can best utilize these tapes are current awareness services or selective dissemination of information services. (See further Section 4.4.) These tapes are less suitable for retrospective searching; that is, the tracing of literature on a given subject over a number of years.

The situation in this field is changing very quickly; but to give an impression of the magnetic tapes which are available already we give the following selection (see Molster, 1975).

Name of tape	Corresponding printed issue
AIM & ARM	Abstracts of Instructional Materials
API	API abstracts of refining litt.
BA Previews, BIORI	Biological Abstracts Biological Research Index
CAC/IC	Current Abstracts of Chemistry and Index Chemicus
CA Condensates	Chemical Abstracts
CAIN	Bibliography of Agriculture
CBAC	Chemical Biological Activities
CLAIMS	Uniterm Index to Chemical Patents (USA)
CMA	Chemical Market Abstracts
COMPENDEX	Engineering Index
Drugdoc	Drug Literature Index
EMA	Electronics and Equipment Market Abstracts
ENDS (Eur. Nuclear Doc. Service)	—
ERIC	Research in Education
Excerpta Medica	Excerpta Medica
F & S	F & S Index
FSTA	Food Science and Technology Abstracts
GEO-REF	Bibliography and Index of Geology
GRA	Government Report Announcements
IAA (NASA)	International Aerospace Abstracts
IDC	Among others: Chemical Registry Files (Chemical Abstracts), Section B, C, and E of the Central Patent Index and The Index Chemicus Registry System (ISI)
INFORM	Abstracted Business Info.
INIS	Atomindex
INSPEC	Physics Abstracts, Computer and Control Abstracts, Electronical and Electronics Abstracts
Int. Labour Doc.	—
ISMEC	ISMEC Bulletin (INSPEC)
I.V.	Index Veterinarius
Mass Spectrometry Bull.	Mass Spectrometry Bulletin
MEDLARS/MEDLINE	Index Medicus
METADEX	Metals Abstracts Index
NSA	Nuclear Science Abstracts

Name of tape (contd)	Corresponding printed issue (contd)
PASCAL	Bulletin Signalétique
Pollution	Pollution Abstracts
POST	Polymer Science and Technology
Psychological Abstracts	Psychological Abstracts
SCI	Science Citation Index (Current Contents)
SSCI	Social Science Citation Index (Current Contents)
SSIE (Smithonian Science Information Exchange)	(Current research, mainly USA)
STAR (NASA)	Scientific and Technical Aerospace Reports
Vet. Bull.	Veterinary Bulletin

An entirely different form of documentation work is the recording of numeric data and not the subject content of the publications. These form 'data banks'. Here too a computer is often used not only to store the information but to arrange it and to provide information. Examples are details of patients in hospitals, legislation, structural formulae of chemical substances and physical constants.

Other documentation services again, commonly linked with special libraries, record information from documents in systems for their own use or for a limited circle of users (for instance within a firm or an organization). They are often card systems of small size or partly mechanized systems with edge punched cards or transparent feature punch cards (e.g. Peek-a-Boo cards). Such systems must be structured very specifically to the information needs of the (limited) circle of users and must therefore by definition be selective.

3.3 Information services

We will now examine another function of documentation services: the giving of information. If this information function comes to the fore, the service is perhaps called a documentation and *information* service, or, briefly, an information service, or information analysis centre.

An information service works differently from a documentation service. In the documentation service, the primary task is the systematic recording of information; in the information service, it is retrieval. An information service will therefore restrict recording to a minimum to satisfy needs not satisfied by other services. It will try to obtain as much documentation material as possible from elsewhere. The staff of such a service must be skilled in operating with diverse documentation systems.

Information provided to clients can be divided into four categories.

Quick information

This is information, for instance, on the name and address of a journal, the composition of a chemical substance, details of a few reference books on a subject, or a few statistical data. Such information may be given by telephone, telex or letter. At the present stage of development of machine systems, there is little purpose in acquiring this sort of information by machine.

Bibliographies

These are lists of bibliographic descriptions from literature on the subject required. The list is specially compiled for the client but since the same subject may recur, these bibliographies are themselves stored in a systematic way by the information service so that they can be used again, so limiting searching work. Such lists may be exchanged between information services.

When a request is made for such a bibliography, the subject should be defined very carefully and information should be given about the circumstances causing the enquiry, the period for which literature is required, and the languages in which literature is required.

Some services make bibliographies with or without abstracts on their own initiative and publish them commercially for a limited group of users. Although such lists can be prepared mechanically, so far it is usually cheaper and more efficient to find the information by conventional means.

Literature surveys or reviews of the literature

Literature surveys are the most elaborate way in which information can be supplied about the literature. Before a start is made on a literature survey, the request must first be discussed, preferably personally with the client. An initial check may be made of what secondary sources are available before further discussion about how to tackle the survey.

The simplest form of survey is that in which abstracts are added to the bibliographic data of the publications (an extension of bibliographies). Then there is the descriptive literature survey in which the information is elaborated into a continuous story. The most elaborate form is the critical subject review, which takes into account the purpose of the specific enquiry.

The literature survey is dependent on intellectual work more than on mechanical processing.

Current awareness

A current awareness service alerts the client to current information appearing on his subject of interest. This function of the information service is the obvious choice for mechanization. For different types of current awareness service, see Section 4.4.

Further reading

Boyle, P. J. & Buntrock, H. (1973). *Survey of the World Agricultural Documentation Services.* (FAO-DC-AGRIS—6; EUR—4680-1e), 300FB.

Europa Publications Ltd. (1972). *The World of Learning 1972–73,* 23rd edn, (2 vols). Europa Publications Ltd.; London.

Molster, H. C. (1975). *Een keuze beschikbare databestanden in computer-leesbare vorm voor bibliografische documentatie* (A selection from available data bases in computer-readable form for bibliographical documentation), pp. 259–64.

UNESCO (1971). *UNISIST Study Report on the Feasibility of a World Science Information System.* (UNESCO—SC70-D-75A). See esp. chap. 4: Functions and trends of present information services.

UNESCO (1975). *World Guide to Technical Information and Documentation Services,* 2nd edn.

4 Primary publications and bibliographic sources

Section 2.4 stated that a literature survey should be based as far as possible on primary publications. There is therefore a tendency to begin literature searches directly with this sort of publication. In other words, the searcher begins with a report or journal article that he obtains by pure chance and uses this publication on a specific topic as a starting point for further study of the literature. Although it is indeed possible to collect literature in this way as discussed in Section 5.1 (the 'snowball system'), a literature survey should begin with the secondary publications and especially with bibliographic sources.

But before dealing with these methods, it is necessary to discuss the concepts 'primary publication' and 'bibliographic source'.

4.1 Primary publications

Primary publications are publications in which the author *for the first time* supplies evidence, describes a discovery, makes or derives a new proposition, or brings forward new evidence about previous propositions. UNESCO (1968) defines primary publication as: 'Original scientific paper, describing new research, techniques or apparatus'.

One cannot be certain that a given publication does indeed describe *new* facts. In general the author must be taken on trust. However attention should be paid to:
  the trustworthiness of the institution where the author works or of the author himself;
  the respectability of the journal or series in which the publication is included or—if a separate publication (book)—the good name of the publisher.
One assumes, under these circumstances, that the manuscript passed through the editorial board before it was published. Institution, journal and publisher share responsibility and must, like the author

himself, maintain their good name. But even an author in good faith can present as new, facts that are not new, because he is not aware of what has been published elsewhere. This can happen, for instance, because the discoveries elsewhere have been published in an inaccessible language or in an unknown (obscure) journal or because he has not studied the literature carefully enough. For this reason alone, every scientific author has the duty of searching thoroughly to find out what is known about the subject before he presents a paper for publication (and even before he begins research). Of course the author's seniors in his place of work, the editors of the journal and the publisher will watch out for this problem, but they cannot make a complete study of the literature relating to each manuscript before providing the fiat for publication. Patent organizations are particularly critical about the novelty of facts. The checking of the originality of patents is an extensive and individual field of critical searching of sources. However, we shall not discuss them further. Likewise, when one is preparing a literature survey or review, it is necessary to be critical about the content of the primary publications used. The bibliography of a primary publication is a good means of checking the character of the publication in addition to a critical assessment of the content of the publication as such. The completeness of the bibliography depends on the nature of the research and also on the character of the scientist. But this bibliography must never be one-sided (dealing only with literature in one language or with publications of a certain 'school' or institute). If one has the impression that the bibliography of a work is indeed biased, one must be extremely careful in accepting the originality of the primary publication. It is very possible that others have priority over the author.

The same applies to the time scale of the literature in the bibliography. If, for instance, a primary publication of 1972 contains references only to a few reference books published prior to 1940 and a few articles published not later than 1957 (thus at least 17 years old), take care. It is of course possible that nothing has been done on this subject since 1957 but it is highly improbable.

A primary publication and in particular a research report must fulfil three requirements:
(*i*) the author must have made it clear in the publication that the research described has yielded new facts (negative facts too can be new!);
(*ii*) the publication must include justification for the research;
(*iii*) the publication should be a condensation of how the experiment ran.

In discussing research reports, we will go further into these points (Section 8.3).

4.2 Bibliographical sources

Primary publications are by nature strongly informative. But they cannot serve as the sole source of information because collectively they are not systematically arranged and are therefore difficult to handle. Hence, alongside primary publications other types of publication have arisen that are easier to handle. The purpose of these types of publication is to condense to a large extent the information obtained and to provide another means of access to it.

These publications can be assembled under the term secondary publications. They are derived from the primary publications. Secondary publications in which an attempt is made in some way to provide a view of existing *literature* are called bibliographic sources. Other sources that are secondary are, for instance, encyclopaedias, reference books, handbooks, monographs and annual reports. The division between primary and secondary publications is not sharp. Some papers presented in the proceedings of congresses may have more the character of reviews, whereas others may concentrate on reporting new facts. There are different types of bibliographic source including the following:

   guides to the literature
   indexes
   abstract journals
   bibliographies in a strict sense
   bibliographies of bibliographies
   literature surveys
   review articles
   progress reports
   catalogues
   documentation systems in card form
   systems for use in computers (magnetic tapes, disc memories, data cells).

4.3 Survey of primary and secondary publications

Here we will discuss the special characteristics of some types of publication. It must first be said that there are so many transitions from one type to another that these characteristics can be only a rough approximation. For example:

Scientific journals are characterized by the primary articles that appear in them. But there are many scientific journals in which more or less of the content is taken up by abstracts or review articles (thus secondary literature).

There is an essential difference between reference books and textbooks. But many reference books are called textbooks. The English term handbook covers a different range of concepts from the Handbuch (German), handboek (Dutch), and manuel (French).

4.3.1 Scientific journals

We can certainly say that the scientific journals—or more broadly scientific periodicals—are still the most important source of information for science.

A characteristic of the journal is the way it appears at set intervals of time (from weekly to quarterly or three times a year). Usually it is published regularly but its regularity often leaves much to be desired. A stated date on a journal of June 1972 is no evidence that the issue indeed appeared in June 1972!

Another characteristic is that an issue usually contains more than one article, although sometimes an issue contains only one treatise (for instance a thesis). The difference between journal and series (consisting of separate issues with one treatise) is therefore not sharp, although a series is not usually periodic. There are therefore many libraries which combine journals and series into one category in the library under the term 'periodicals', in which the criterion of a periodical is that it is an unlimited series of publications. An encyclopaedia published serially is therefore no periodical. As to whether annual reports or proceedings of congresses held regularly, and thus forming an unlimited series, can be included in this group is an open question. A third characteristic of scientific journals is that they have an editorial board that ensures that the content of the journal meets the aims of the journal and that for instance the content of manuscripts submitted are critically assessed. Sometimes the editorial board does not examine them itself but uses the assistance of a number of outstanding specialists (called referees) who, either anonymously or not, assist in this assessment.

In a normal scientific journal the following elements can usually be found:

on the front cover is the name of the journal, the volume number or year, the number in the volume (or issue number), and the date on which the issue appeared;

the list of contents (on the outside or inside of the front cover; sometimes on the back cover or elsewhere);

list of members of the editorial board, the address of the editor or of the administration, the annual subscription (all these are often on the inside of the front cover);

notices to contributors on how manuscripts should be submitted and conventions about how literature should be cited, form of bibliographies, tables and figures (usually on the inside of the back cover);

the actual contents consisting of articles;

the page numbers at the top or bottom of each page but often not on the first page of an article. As a rule the page numbers are in one sequence for a whole volume;

a running head at the top of each page (except the first page of an article) with the name of the author on the left page and an abridged title of the article on the right page;

at the foot of each page a bibliographic strip giving the (abbreviated) name of the journal, date, volume, issue number and page number;

sometimes a special sheet or page with abstracts of the articles;

sometimes an introductory article in each number written by or on behalf of the editorial board (editorial);

often book reviews, abstracts, society news, news of members, brief scientific news (letters to the editor, short communications) and submitted articles (whose subject matter is not the responsibility of the editors);

often advertisements.

An article in a journal includes the following elements:

the title of the article;

the name(s) of the author(s);

the address(es) of the author(s) (usually name of the institution and place where the author works);

sometimes (as footnote) a brief biography (or biographic details) about the author;

a summary (annotation, indicative or informative abstract or author's abstract, see Section 6.3); usually now placed at the beginning of the article;

the content of the article, divided by headings;

the bibliography (list of cited publications);

sometimes appendices.

The oldest scientific journal is the French *Journal des Sçavans*, of which the first issue appeared on 5 January 1665. It was in fact the continuation of scientific correspondence that then existed and took the form of news. It covered all the sciences. This initiative was followed three months later by the *Philosophical Transactions of the Royal Society* in London whose content formed the prototype of present scientific journals: scientific transactions. But this journal, too, included all sciences.

Since then the number of journals and the volume of their contents has increased gradually in what is calculated to be a logarithmic curve. The number of current primary scientific journals is estimated to be 40 000, but this estimate is probably low. In total, these journals have produced 6 million articles and the number will now increase by at least $1\frac{1}{2}$ million per year. This line cannot of course be extrapolated and it is envisaged that the increase in the number of journals and journal articles will slow down in the near future. In other words the scientific journal will have passed its optimum as a means of communication.

There are indeed symptoms that reflect this. Many scientific journals live from hand to mouth and are forced into accepting subsidies and into the institution of page charges for articles published. In other words, they force the authors (or the institutions where the author works) to contribute towards the cost of publishing the journal. This threatens the independence of the journal as an objective means of scientific communication.

Alongside the journal, other means of communication have arisen, for instance series of separate reports which compete with the journal. Machine storage of scientific data in computer memories is another development that may compete with the classical journal.

Attempts have been made to make a new structure for the journal, for instance separate articles bundled together according to the needs of readers. But these attempts have not had much success. See for instance Phelphs & Herlin (1966). Despite such developments journals compete strongly with books. In general the scientist prefers to publish his results in journals rather than books. Through journals, he can gain an earlier priority date for his results than through books. Books are being used more for stating the position of science at a given time.

We can classify primary scientific journals by two criteria: organization and content.

Organizational journals can be divided into the non-commercial and the commercial. The first category includes many of the journals published by learned societies, universities, laboratories and institutes. There is often a bond between membership of an organization and subscription to a journal: by paying the contribution, the journal is provided free. The editorial preparation is not professional. This work is unpaid or paid by a minimal honorarium despite the fact that the editor or editorial secretary may sometimes devote more than a third of his working time to editing: a sort of hidden subsidy.

Not directly commercial (but certainly indirectly!) are the scientific journals published as house journals of commercial companies. These are usually professionally edited. Although of course the content of these publications can to some degree be biased (they would not usually publish something that conflicted with the concern's interests), these journals still often contain interesting scientific news.

Commercial journals are published by commercial publishers. The purpose is of course to make profits. But many commercial publishers—known, not entirely correctly, in the trade as scientific publishers—are prepared to publish scientific journals at minimum profit or even at a loss. One of their purposes in this is the good liaison they thereby establish with scientific authors and also they gain more publicity as scientific publishers. The preparation of manuscripts for the printer and the commercial exploitation of these journals is of course professional. Usually the publisher requires some assurance that the editorial board keeps the content of the journal above criticism.

A transition between commercial and non-commercial journals are those under the auspices of learned societies or institutions but prepared for the printer by commercial publishers. Sometimes special bureaux have been formed to prepare work for the press (for instance Pudoc at Wageningen, Netherlands).

By *content* primary scientific journals can be classified into archive journals and reading journals. Archive journals are those consulted mainly during retrospective searching of the literature. In the main, they contain research reports. They are directed to the needs of the scientist–authors.

Increasing specialization has brought the tendency for more general archive journals to split into more specialized ones. As specialization increases, the circle of readers diminishes so that exploitation has become more difficult. Just because of this specialization, there has been an increase in the need for journals which keep readers up to date with the *general* progress of science. This type of journal has been encouraged by the increasing confluence between areas of science (interdependence). Journals intended for general scientific readers contain shorter comment and also review articles.

A fairly new category of journal for reading is that dealing with the principles of science (organization, manner of thoughts, social involvement, ethics); the science of sciences.

To find the primary scientific journals that should be consulted for a certain subject one can best begin with the journal catalogue of the library. With further searching of the literature, one fairly automatically finds other journals important for the particular subject. Once one knows the most important abstract journals for a subject, one can usually find a list of the journals covered by that service (at the beginning or end of a volume). There are also general international catalogues such as the *World List* (1963–1965) and *Ulrich's International Periodicals Directory*, 15th edn. (1973). Alongside them there are numerous bibliographies of journals on particular subjects and lists of journal collections in libraries.

4.3.2 Trade journals

Although many journals are published by technical associations, there are also innumerable trade journals produced and published commercially. There is a continuous range of transitions from primary scientific journals to trade journals. Many trade journals also contain primary articles on applied research but alongside them are articles paying more attention to applications of research published elsewhere, to reviews of technical problems, to society news, news of members and such like.

Particular attention is usually paid to typography: the editors seek an attractive form; the format is often larger than that of scientific journals and there is a conscious effort to make them readable. The editor must have a good nose for news that will interest his readers and he must have more journalistic background than the editor of a scientific journal. The reading of trade journals in a particular subject is important especially for those in applied research and development in order to maintain contact with questions that affect the commercial sphere. Trade journals are usually national—or even local—and are written in the native language of intended readers. The number of copies issued varies widely but can sometimes run into hundreds of thousands. They are often financed by income from advertisements. Authors often receive an honorarium in contrast to the authors who publish in scientific journals.

4.3.3 Research reports, other reports and theses

The publishing of research reports printed in book form as separate reports in series is on the decline. This manner of scientific communication is becoming too expensive. Even so there is a tendency to keep to this form, particularly by academic libraries which can use such monographic research reports for exchange with other libraries.

This type of report in book form is maintained too by universities, laboratories and institutes financed by the government. The collective name of the series is commonly Report of ..., Contributions to ..., Verslagen van ..., Mitteilungen. ... Each publication in the series is numbered, sometimes in sequence, sometimes by annual volume. This last type approaches the form of the journal, but with only one article per issue and with irregular issue of the numbers. Often they have an annual list of contents or index.

It is particularly confusing that these series often contain reprints or offprints of articles from journals. One then has the impression that the publication is new when it is in fact a reprint of something published elsewhere. This impression may be strengthened by the careless manner in which reprints are put out: the original place of publication may be inadequately referred to if mentioned at all, the pagination may be changed, pieces may be added or taken out and sometimes the whole structure may be so altered that parts of the text appear on different pages from the original text.

During and since World War II, another type of report has come to the fore: the report reproduced in a simple manner in small numbers. There are various reasons.

In the War, much research was done for military purposes. It was of course necessary to give the results of this research in reports but they were 'classified' or had to be kept secret. They were reproduced in small numbers and distributed only internally. The form was simple: stencil or offset or even photocopy. After the War, many military scientific reports were released or 'declassified' and so gave rise to a new stream, or even flood, of scientific information in the form of declassified reports.

A second reason is that many good scientific journals receive too much copy and cannot publish it within a reasonable period. There is therefore a time lag in publication of up to a year or more, which scientist–authors find unacceptable. In the competition between scientists, people try to report their results and publish them as quickly as possible so that they have priority in their discoveries. A simply reproduced report, distributed from one's own institute to colleagues, is ideal for this purpose.

A third reason is increasing specialization. The circle of colleagues interested in a particular piece of research is decreasing; scientists know their colleagues personally and nothing is simpler than to keep them informed by a report. De Solla Price has called such groups 'invisible colleges', a sort of closed community with their own language and exchange of information. In this way, the separate report is taking over the function of the reprint from the scientific journal.

A fourth reason is probably that the report described here can be prepared more quickly (but less carefully) than the journal article. One is less worried by the criticism of editor or editorial board and one is less restricted by numerous regulations and standards set by the editors of journals. A sketch for example on graph paper may suffice; one does not have to use a qualified draftsman to prepare illustrations.

A final reason is the improvement in methods of reproduction, in particular of offset which is no longer considered to be a poor relation of letterpress.

Obviously this uncontrolled and chaotic stream of reports reduces the librarian and documentalist and those concerned seriously with literature searching to despair. The more so, since alongside this stream of 'official' (in other words authorized or public) reports, is a completely invisible stream of internal reports within firms and companies for which the secrecy is broken from time to time by publication or declassification.

There is some attempt to order this situation by systematic storage and dissemination of at least some of those reports of research paid for by the government or international organizations. Examples are the National Technical Information Service (NTIS), the National Aeronautics and Space Administration (NASA) and the Atomic Energy Commission (AEC) in the United States, the International Atomic Energy Agency in Vienna (IAEA), the Food and Agriculture Organization of the United Nations (FAO) in Rome, the British Lending Library (BLL) at Boston Spa, Yorkshire. The NTIS at Springfield (Va) yearly adds 40 000 titles to its collection and issues an abstract journal twice a month entitled *US Government Research and Development Reports* with an index. All reports can be ordered through this bureau in photocopy or microform. Subject by subject, they also publish *NTIS Announcements in Science and Technology*. Ad hoc information on a subject can be supplied too. NASA publishes *Scientific and Technical Aerospace Reports*, which survey the international report literature, AEC produces an abstract journal twice a month: *Nuclear Science Abstracts*. It works closely with the International Atomic Energy Agency (IAEA) in Vienna where the International Nuclear Information Service (INIS) makes the world literature in this field accessible. The FAO publishes *Agrindex* monthly, a current title information service on agricultural science. In the United Kingdom, the British Lending Library at Boston Spa functions like a clearing-house. This library publishes a monthly *British Research and Development Reports*. These reports are likewise obtainable in photocopy or microfiche from that address. Twice a year they publish *R & D Abstracts*.

Dissertations and theses too are a difficult source of information. In a few countries, like the Netherlands and the Federal Republic of Germany, it is customary to print a thesis. In other countries where they are not printed, they correspond more to report literature.

One tends to consider theses only as primary literature. The purpose of a thesis is to demonstrate that the student can carry out research and write a report on it. This is indeed true but it is customary to include an extensive literature survey in the thesis. This part (and the bibliography) can serve as a secondary source in searching for literature.

The cost of printing a thesis is high for the student, especially as it usually contains 80–100 pages. Ways have therefore been sought of reducing these costs. As well as giving subsidies, some university libraries commonly buy copies of the thesis for exchange. Another arrangement is for the thesis to be included in a series published by the university, and also to print a limited edition as the true thesis. Now and then journals can publish theses by means of subsidies either as serial articles or as monograph 'supplements'. However, journals are not the best medium for publishing such bulky works. Such double publication can lead to confusion. There must be clear cross-reference from the one type of publication to the other. A thesis can rarely be published commercially.

For a true thesis, there is always some statement on the title page such as 'Doctoral thesis or doctoral dissertation for ...'. In some countries the form of the thesis has recently been in the melting pot. The regulation that the thesis be printed in letterpress has been rescinded. Other means of reproduction (offset or even stencil) are now allowed. More and more theses are issued in a foreign language (especially English) even though, because of the law, the title page must contain the statement that it is a thesis in the official language of the country.

Secondly—and this is more significant—the content of the thesis is under discussion. There is a tendency to drop the requirement that the thesis be a single complete publication. Graduation on (already published) journal articles, perhaps supplemented with a short narrative and with reference to these articles, is already in use. An ideal form has yet to be found. The most important bibliographic source for dissertations is undoubtedly *Dissertation Abstracts International* (address: University Microfilms, P.O. Box 1764, Ann Arbor, Michigan 48100, USA). Since 1970, this service has included not only American dissertations but also European. The dissertations they list are supplied in the form of microfilms.

There are also many national bibliographies of theses. For Great Britain there is *Index to Theses, accepted for Higher Degrees by the Universities of Great Britain and Ireland and the Council for National Academic Awards* (Aslib, London). For Germany, *Jahresverzeichnis der Hochschulschriften—DDR, BRD und West Berlins* (VEB Verlag für Buch- und Bibliothekswesen, Leipzig). For France, there are various series including *Catalogue des Thèses de Doctorat soutenus devant les Universités Françaises* (Direction des Bibliothèques et de la Lecture Publique, Cercle de la Librairie, Paris). In some other countries theses are included in the national bibliographies. In the Netherlands the library of Utrecht University publishes an annual *Catalogus van Academische Geschriften* (Nieuwe reeks), which includes the theses of all universities. The journal *Higher Education and Research in the Netherlands* (NUFFIC, The Hague) also contains a list of Dutch theses in each number.

4.3.4 Proceedings of congresses

Each year congresses and symposia are held in all countries of the world. They are national, multinational and international. Some congresses have thousands of participants and others are symposia for a select group of experts. There are congresses with a clear political motive—UNCTAD at Nairobi, Kenya (1976), the Congress on the Human Environment at Stockholm (1972), the World Food Congress in the The Hague (1970)—and congresses that fulminate less and are confined to scientific matters. The papers of these congresses are usually published, but the form in which they are published varies widely. The first purpose of most congresses is for colleagues in a subject to meet and exchange ideas.

This does not mean that proceedings are not important enough to be considered as a source of information for literature study. Usually the outstanding workers in a field are invited as speakers at congresses so that the content of the proceedings gives a good indication of the state of affairs at a given moment.

Some papers are state-of-the-art reviews while others present new material. The proceedings of congresses lie therefore between primary and secondary literature. As well as invited lectures, there is often occasion for free contributions. Since the quality of these free contributions often leaves much to be desired, the consultation of proceedings of congresses is sometimes tiresome. Discussions, if they are included, may be more of a hindrance than a help.

Unfortunately the publishing of proceedings often runs into difficulties, particularly if the report must be published after completion of the congress. Despite all precautions, the editors can seldom lay their hands on all the texts to be published at the close of the congress. The process of gathering together scripts, once the participants have departed, wastes much time. Hence proceedings often appear several years after the congress was held and the factual content is no longer topical.

Often the publishing of proceedings is financially unattractive; congress organizers pay little attention to this. Sometimes the contributions may be distributed as preprints before the congress; sometimes there are both preprints and a book afterwards. Otherwise the contributions, or at least some of them, may be published separately in a journal (in a special congress number or spread throughout several numbers). It can happen that contributions are published in different journals.

Usually the host country is responsible for publishing the proceedings. This again does not help to make proceedings accessible: for one series of congresses, it is necessary to chase up each publisher or congress organizing committee. There is little uniformity in the manner of presentation.

Another bad result of proceedings is that the separate papers are less often abstracted in abstract journals than are articles from journals. Most material is thereby lost to posterity. It has been observed that proceedings of congresses are less often cited by scientists than are other forms of literature.

Even so, it is advisable when starting a literature survey to trace proceedings dealing with the subject. It is an easy way of getting to know the state of the art. The simplest way of tracing proceedings is through the systematic (classified) catalogue of the library. Sometimes the library has a special catalogue of congress proceedings.

There are numerous guides to proceedings or at least to congresses due to be held, such as the *International Congress Calendar* (Union of International Associations, Brussels), *Forthcoming Scientific and Technical Conferences* (Dept of Education and Science, London), *Scientific Meetings* (Libraries Association, New York),

World List of Future International Meetings (Library of Congress, Washington), *World Meetings* (TMIS, Newton Centre, Mass.) and *Aslib Information* (monthly; Aslib, 3 Belgrave Square, London SW1X 8PL). Directly relevant to proceedings of congresses are the bibliographic periodicals: *Bibliographical Current List of Papers, Reports and Proceedings of International Meetings* (Union of International Associations, Brussels) (monthly); *Directory of Published Proceedings* (InterDok, White Plains, N.Y.) (monthly); *Proceedings in Print* (Mattapan, Mass.) (bimonthly); *Index of Conference Proceedings received by BLL* (British Lending Library, Boston Spa, Yorkshire). An index of separate papers is *Current Index to Conference Papers in Chemistry, Engineering, Life Sciences* (CCM Information Corporation, New York). The European Commission at Brussels is planning a special documentation service on papers of congresses and symposia on environmental hygiene.

4.3.5 Reference works, treatises and monographs

Reference works and monographs have a distinct role. There is no principle difference between them. Webster's dictionary defines handbook as: (*a*) a book capable of being conveniently carried as a ready reference (syn. manual); (*b*) a concise reference book covering a particular subject or field of knowledge; monograph is defined by the American Heritage Dictionary as a scholarly book, article or pamphlet on a specific and usually limited subject. The difference is quantitative rather than qualitative.

The influence of reference works as a source of information has declined. The compilation of a reference work on a subject—even a limited subject—is becoming almost impossible. One has to work with different authors and the whole operation is so extensive and time-consuming that the work is obsolete before it appears. In consulting reference works, one must therefore look for the date on which the separate chapters were closed by the authors (and not the date of publication given on the title page).

There is a tendency for large old-fashioned reference books to be replaced by monographs, books of more limited scope (often paperback) on more specialized subjects. Monographs are not very different from reviews. There is rather a continuum in that monographs are based on the author's knowledge; whereas reviews are based more on the literature.

A reference book thus treats a fairly large or large area of science systematically. In this, it differs from a textbook which must have a didactic approach. To put his material across, the author of a textbook should preferably begin with a simple (and often practical) example to demonstrate what he means. He completes and extends his treatment by further examples and can then present general laws or theories. For the reference book, systematic treatment is central and examples are given only as illustration. In consulting a reference book, one must be past the learning stage to the studying stage, in other words the gaining of understanding.

The term 'Handbuch' (German) or 'handboek' (Dutch) is often confused with the English word 'handbook', which is often used in an entirely different meaning from 'reference work'. It refers also to a book that is constantly to hand on one's desk or laboratory table. This concept has nothing to do with the 'handbook' containing a treatise but is rather synonymous with the word 'manual'. Because of its systematic treatment, the reference work must contain a thorough subject index. The book must be usable for reference as well as for study.

In a first approach to the literature of a subject, reference works cannot be missed; in deeper study, this source is less useful. The reference work never goes deep enough into a specialist subject; it will never deal with the latest literature on the subject. The monograph is a far better means of entry.

Some types of reference work contain extensive tables or tabular surveys. They are essential for laboratory work. In literature searching, these reference works will be used only for tracing basic data. Reference works and monographs contain secondary information. The authors or editors always attempt to include the latest information in these publications but not to publish primary information in them. In the past this was indeed so. Reference works, treatises and monographs are usually published by commercial publishers.

4.3.6 Encyclopaedias

Encyclopaedias can give a first easy introduction to a subject. They are commonly alphabetically arranged but often contain more keywords than entries, some being traced by cross-references. A few encyclopaedias are arranged systematically. In this type, the main chapters deal with the most important subject areas.

In compiling the list of words to be included in an encyclopaedia, use is nowadays made of a computer so that the encyclopaedia can have an extensive subject index (usually as final volume). It is advisable to consult this subject index because it contains many more points of entry than the text of the encyclopaedia.

The preparation of multi-volume encyclopaedias takes years so that the first volumes are often partially obsolete before the last volumes have been published. The difficulty of obsolescence is to some extent circumvented by yearbooks or supplements to the encyclopaedia.

The advantage of an encyclopaedia is that the composition and compilation are put in the hands of an editorial board, advisers and specialists with deep knowledge and experience; the aim is brevity but clarity. Each term that is included is defined and explained. These definitions, or at least descriptions, are often useful for literature surveys.

For science one should rather refer to the large general encyclopaedias than the more popular encyclopaedias designed for a wide public. Each library of any consequence possesses one or more general encyclopaedias. Besides these general comprehensive encyclopaedias, there are many specialized ones. These can be found in guides to the literature and in the libraries themselves. Libraries place them in the reading room where they can be consulted, but do not, of course, lend out these reference works.

4.3.7 Guides to the literature

A scientist should be familiar with the bibliographic sources for his subject field. Besides the research literature that he reads to keep up with his subject, he must constantly refer to these bibliographic sources to find out what he needs to know about specific subjects he is working on (retrospective searching). But for the student it is no simple task to find which bibliographic sources are important to him. It is made less easy by his need to deal with a number of disciplines.

Luckily there are guides that ease the approach to bibliographic sources and to literature on a given discipline. A book that provides an approach to these guides to the literature is Grogan (1970). This book deals with all sorts of publications pertinent to science and for each sort describes the approach to this sort of publication. There are also more specific guides to the literature that could be called manuals to the literature on particular disciplines. Grogan describes many of these guides; here a few will suffice (under 'References' many more titles are listed).

General guides are Jenkins (1965) and Fleming (1957).

For the chemist, there are various: Melton (1965); Burman (1966); American Chemical Society (1961); Bottle (1969); Crane *et al.* (1957); Dyson (1958).

The physicist can use Whitford (1954); Yates (1965); Parke (1959) and Weiser (1972).

Important guides for biologists are Bottle & Wyatt (1971); Smith and Painter (1966) and Troyer *et al.* (1972).

Guides to technical subjects include Gibson & Tapia (1965) on metallurgy; Pritchard (1969) on computers; Malinowsky (1967) on science and engineering; Carey (1966) on technical information; Herner (1968) on science and engineering; Fry & Mohrhardt (1963) on space and science technology.

4.3.8 Indexes, abstract journals and bibliographies

The simplest way of providing an access to the literature is an index. In other words, the titles of publications are arranged in such a manner that the user can use this list to trace publications in the list on subjects that interest him.

Two aspects of this indexing are particularly important: the number of publications indexed and the indexing system. The two aspects are related. As the number of publications in the index increases, it is necessary that less intellectual work per title goes into the indexing.

An example for explanation: in general a volume of a scientific journal contains an annual index. The compiling of this index will usually be done by the editor or editorial secretary and will take him one or two evenings. But if one proposes to index the agricultural literature which is estimated to contain yearly 200 000 or 250 000 publications, an entirely different problem is posed. In such a system intellectual work for bringing each publication into the system must be restricted to the minimum. Clearly the system cannot be perfect, especially since these indexes must be as recent as possible and time must not be lost in compiling them.

The user must remember that these large indexes are useful for a first superficial approach to the literature of the subject he wants to study. He will find much literature in these indexes that on further examination proves irrelevant.

Of course, there have been attempts to automate the production of these indexes. The huge administrative work of preparing these was ideally suited to the computer. Consequently, alongside the printed publication—by computer and photo-typesetting—other products have become available to users: magnetic tapes with the same data are now an attractive form. The user can obtain information mechanically from these magnetic tapes with his own computer centre. This approach is especially attractive for current awareness or selective dissemination of information.

The way in which the printed indexes are prepared is very variable. Before looking up any of these indexes, one must first learn which system has been used. Usually there is an explanation at the beginning of each issue. Sometimes the index is divided into broad categories, which are subdivided by keywords. Usually, and certainly with the large indexes like *Bibliography of Agriculture*, this is not enough. The titles of publications are indeed put together into categories, but a further index in each issue of the bibliography allows deeper retrieval of information from the literature. Under each keyword of this index, there are a number of words from each title that indicate the content of the publication as far as possible. For other indexes, there may be a fixed list of keywords (a vocabularium or thesaurus) that contains thousands or even ten-thousands of keywords. After the keywords are the references (by numbers) to titles coming under the keyword.

Another way is to place each significant word from the title in the middle of a page with the rest of the title around it. In this way one obtains an alphabetical list in the middle of the page in which one can search. As explanation an example:

We start with the following four titles:

'The need for documentation to government specifications';

'Bibliographical style manuals: a guide to their use in documentation and research';

'Integration of technical editing with other documentation functions';

'The documentation program of the Special Libraries Association'.

Under the word 'documentation' in the middle of the page of the index we find:

```
                       THE  NEED   FOR  DOCUMENTATION  TO  GOVERNMENT SPECIFICATIONS.         A-1112
LIOGRAPHICAL STYLE MANUALS—A GUIDE  TO THEIR USE IN  DOCUMENTATION  AND   RESEARCH.           BIB B-1770
         INTEGRATION  OF  TECHNICAL EDITING  WITH OTHER  DOCUMENTATION FUNCTIONS.             A-1118
CIATION.                                          THE  DOCUMENTATION PROGRAM  OF THE SPECIAL LIBRARIES ASSO A-0341
```

In the same way we find the first title also under GOVERNMENT, the second title under STYLE MANUAL, the third title under TECHNICAL EDITING and the fourth title under LIBRARIES. Sometimes keywords from the title are supplemented by extra keywords.

Naturally, the end of each line of the index contains an indication of the place where complete details of the publication can be found in the index itself. Many documentation services prepare this sort of index on large fields of the international scientific literature. The best known is probably the American MEDLARS (US National Library of Medicine, Bethesda) that has published *Index Medicus* since 1964. More than 13 000 titles are processed per month. MEDLARS is also available on magnetic tape. *Excerpta Medica* (Amsterdam) runs a similar service.

Bibliography of Agriculture contains over 100 000 titles a year. This index is available on magnetic tape under the title BoA tape (Bibliography of Agriculture tape). Another tape on the same subject is the CAIN tape (CAtaloging and INdexing) prepared by the National Agricultural Library (Beltsville, USA). The United Nations Food and Agriculture Organization publishes various indexes, which cover reports prepared for FAO as well as statistical data.

In 1975, under the auspices of FAO, an international system for retrieval of the literature on all subjects of FAO (including food, forestry and fisheries) was started. This system will be divided in two parts: AGRIS I and AGRIS II. AGRIS stands for International Information System for the Agricultural Sciences and Technology.

The object of AGRIS I is to produce a comprehensive index called AGRINDEX, to all literature relevant to agricultural science and technology (200 000–250 000 titles annually). AGRIS II will be a common framework of services producing specialized abstract journals.

Abstracting services also produce indexes for which the tape versions are important. Examples are: *Chemical Titles* and *Chemical Abstracts Condensates* by Chemical Abstracts Service; *Biological Abstracts Previews* by Biological Science Information Service—BIOSIS; *COMPENDEX* and *PANDEX* (Current index to scientific and technical literature) prepared by *Engineering Index Monthly* and *CAB tapes* prepared by Commonwealth Agricultural Bureaux. INSPEC (International Information Services for the Physics and Engineering Communities, London) supplies bibliographic data on tapes covering physics, electrotechnology, computers and process control (see also pp. 22 and 23).

Related to these indexes of the literature there are also indexes of current research based on project descriptions, for instance *Bioresearch Index* (prepared by BIOSIS). These indexes to current research are usually still national. There is, however, an attempt to reach international schemes of co-operation. In the European Community, for instance, an index is being established for current research projects in agricultural and applied fields in countries of the Community. The first experimental issue has been published. This index (AGREP—AGricultural REsearch Projects) will correspond to an American index called CRIS (Current Research Information Service) (Smithsonian Institute, Washington) and an FAO index for developing countries (CARIS: Current Agricultural Research Information Service).

Finally, a special form of index is *Science Citation Index* (Institute for Scientific Information, Philadelphia) in which about 300 000 articles are included per year. To *Science Citation Index* belong *Source Index* and *Permuterm Subject Index*. These services have existed since 1963. *Source Index* lists alphabetically by the name of first author all articles that are examined for *Science Citation Index*. This index is thus an annual author index to a large number of carefully selected journals, chiefly on natural sciences, medicine and technology (engineering). *Science Citation Index* lists alphabetically by name of first author all articles which are cited in the articles of *Source Index*. It is thus an index of *cited* publications. *Permuterm Subject Index* lists all important title words from the *Source Index* alphabetically. It can be used for rapid searching. If one knows a few words from a title, that is enough to find the article itself.

We will come back to this index in describing the 'snowball' system (Section 5.1). The above account shows that indexes do indeed give rapid information about the literature but do not go deeply into textual analysis. The title of a publication does not say everything about its content!

To improve the information, keywords can be added to the title and these keywords too can be indexed, but this demands much intellectual work from specialists and detracts from speed. If one insists on deep indexing + completeness + speed, there are difficulties (financial consequences apart!). Attempts are being made to search for keywords in articles mechanically, but such a system is not yet operational.

Abstracting services publishing abstract journals are based on another principle. Here intellectual work comes to the fore and the aim is not primarily completeness but critical selection. Speed is sacrificed for accurate and thorough analysis of the literature.

Abstract journals attempt to cover the current literature of a given field by bibliographic and factual analysis in shorter or longer abstracts. The abstracts are arranged under subject headings to give a quick systematic view of the literature in each field. These abstracts are usually written by subject specialists and can therefore be expected to meet high standards. Some abstract journals cover very broad fields (for instance *Chemical Abstracts*, *Biological Abstracts* or *Physics Abstracts*); others cover narrow fields (for instance *Weed Abstracts*).

A few points should be considered by the user of abstract journals:
(*i*) There is no uniform system in the indexing of these journals. One must first study thoroughly how the abstracts are arranged by subject headings, categories or sections, or how they can be approached from indexes published in each issue or annually before starting a search.

(*ii*) Studies on the coverage of literature in the abstract journals have shown that their coverage varies from 60 to 80 per cent of the relevant literature. Articles from journals are usually better covered than books; and of the books, particularly weak links are theses and congress proceedings. In tracing literature on a certain subject, it is therefore advisable to consult more than one abstract journal.

(*iii*) There is always a delay in making abstracts. A time-lag of nine months is normal between the appearance of the original article in a journal and the appearance of the abstract in a secondary journal. Sometimes this period may be several years. The latest literature cannot be traced.

(*iv*) An abstractor is only human. Despite the objectivity he aims at, he cannot avoid some subjective influence. This subjectivity is particularly evident in motives for selection.

(*v*) Abstracts are not critical reviews. At most, the abstractor may indicate errors or mistakes by remarks in square brackets. But by reading between the lines, one can get an impression of the value the abstractor places in an article. For instance: 'High temperatures in short days appear to affect . . .' or 'It is suggested that high temperature delays . . .' is much weaker than 'The treatments also increased the rate of . . .'.

In the world, there are over 2000 abstracting or indexing services. They are listed by National Federation of Science Abstracting and Indexing Services (1963), UNESCO (1965) and other directories. These organizations, of course, overlap in their services. A study has shown frightening figures: the number of overlap (in other words the number of times one article is abstracted) is about 1.5. But this value is not really unsatisfactory. One must remember that many articles are relevant to more than one subject and it is therefore not absurd that two abstract journals on different subjects should abstract the same article.

Duplication is also acceptable for different languages. One may not, for instance, expect that Russian scientists should regularly read abstract journals in English or that English scientists should read Russian abstract journals.

A small survey by Pudoc (Halászi, 1968) on the abstracting of Dutch agricultural literature brought the following conclusions. Of 32 Dutch agricultural and biological journals for the year 1962, 85 per cent of the scientific articles and 67 per cent of the semi-scientific articles proved to be abstracted once or more in 35 Western and East European abstract journals. Of these articles, 60 per cent were abstracted more than once. If one excluded *Bibliography of Agriculture* (which gives only titles), only 34 per cent were abstracted twice or more. Duplication in one language occurred for only 11 per cent of the articles. Further, on considering the content of the article and eliminating articles that impinged on more than one discipline, real duplication hardly ever occurred.

Like indexing services, many abstracting services have gone over to the use of computers for production, especially for indexing abstracts. This has considerably improved retrieval. Previously there was often a long delay in publishing indexes and there was only an annual index, whereas now there can be indexes to each issue and an accumulative index at the end of each volume.

It is impossible to enumerate the most important abstracting services for each subject. Details can be found in the directory compiled by the National Federation of Science Abstracting and Indexing Services (1963) and in guides to the literature (Section 4.3.7). Let it suffice to mention a few examples of the large services: *Chemical Abstracts*; *Biological Abstracts*; the 18 abstract journals of Commonwealth Agricultural Bureaux; *Science Abstracts*; *Bulletin signalétique*; and *Referativnyî Zhurnal* (in Russian).

After the survey of indexes and abstract journals, we can deal summarily with bibliographies. Indexes and abstract journals are in fact also bibliographic publications and many carry the title bibliography (for instance *Bibliography of Agriculture*).

Bibliographies are publications in the form of lists of titles with or without notes (annotations or abstracts). They are usually arranged systematically. There are many sorts of bibliography: bibliographies made incidentally on a certain subject or on a certain discipline; bibliographies appearing periodically or with periodic supplements. The distinction between current bibliographies and abstract journals is not sharp, especially since many abstract journals also publish or contain bibliographies. Other publications too can be considered under the category bibliography: a thesis with an extensive literature list; a review article; a reference book; or a monograph.

There are again so many bibliographies that it is necessary to make bibliographies of bibliographies. These are listed in guides to the literature (Section 4.3.7). Alongside tools already mentioned for tracing bibliographies, these bibliographies of bibliographies are very important in starting a search for literature. The authors of bibliographic works are mostly from the library or documentation field. They have therefore a somewhat different approach from the authors of review articles who come from a research background. Librarians and documentalists have tended to work for completeness. In this way, these bibliographies can be better sources than subject reviews. However, their authors are sometimes less critical of the information contained in the publications.

4.3.9 Review articles, literature surveys, progress reports etc.

Review articles are generally written by subject specialists. They commonly contain critical comment. They are invaluable for quickly learning the state of a certain science or subject. Usually they are based on study of the literature but sometimes also summarize the author's own research. UNESCO (1968) defines a subject review article as 'a survey of one particular subject, in which information already published is assembled, analysed and discussed. The scope of the article will depend on the journal for which it is intended. It is the duty of the author of a review article to endeavour to give credit to all published work which has advanced the subject, or which would have advanced it had it not been overlooked.'

Two types can be distinguished: incidental reviews and current reviews. This last category is a type of series. In assessing these reviews, one must particularly examine the bibliography, in particular the languages the author can read. It sometimes happens, for instance, that a French review covers only French literature and that there are many gaps in the treatment of English or Russian literature.

One must also check when the literature search was finished. The date of publication is no criterion. If the date of finishing is not mentioned, one must note the date of the most recent publication in the literature list. Many series of this type have titles beginning with the words 'Advances in ...', 'Progress in ...', 'Survey of progress in ...', 'Yearbook of ...', 'Review of ...'. There are also directories to this sort of publication: UNESCO (1965) and National (British) Lending Library (1964 and 1966). There are also bibliographic publications on these reviews, for instance *Reports on Progress in Physics, Chemical Reviews, Annual Reports on the Progress of Chemistry, Annual Review of Biochemistry, Bibliography of Medical Reviews, Index of Reviews in Organic Chemistry.*

4.3.10 Annual reports

A research institution's annual report usually surveys the state of their research. Much, of course, is not yet published; but it also contains a list of publications issued during the year. Especially when the 'geography of research' (Section 5.2.2) is a basis for the literature search, annual reports are an important aid.

4.3.11 Card systems and other systems

Unfortunately there is such a variety of systems that the whole range cannot be enumerated here. One has to enquire what is available for tracing literature in a particular place.

In literature searching, many people go straight to these (card) systems and never search elsewhere. It seems rather easy and one quickly finds literature. Such a procedure is, however, dangerous. Often these card systems are no more than a random collection of data from the literature. They can be used for a first approach or for a final check but not as the only basis for a search.

4.3.12 Summary

Without attempting completeness, we can now enumerate the following sources of primary and secondary publications.

Primary publications	*Primary/Secondary*	*Secondary publications*
series	monographs	encyclopaedias
reports	proceedings of congresses	reference works
reprints		textbooks
patents	theses and dissertations	guides to the literature
	journals	indexes
		abstract journals
		bibliographies
		reviews of the literature

4.4 Current awareness service

Section 4.3 has sometimes referred to the problem of time, in particular to delays. A disadvantage of all the bibliographic sources mentioned is the time-lag. Even primary publications have a delay. One must allow *at least* half a year between completion of a manuscript and its publication. However, there is an urgent need for *current* information about progress. The publications mentioned so far do not allow this. Other ways have therefore been created and are still being perfected. All these activities are included under the term 'current awareness services'. Below we shall discuss some forms of current awareness.

4.4.1 Short communications

A medium has been created for publishing information more rapidly than by the normal channels of research reports. This is in the form of preliminary publications or preliminary communications in the scientific journals, often under the heading of 'notes', 'letters to the editor' or 'short communications'. UNESCO (1968) regards a text as a provisional communication or preliminary note, 'when it contains one or more novel items of scientific information, but is insufficiently detailed to allow readers to check the said information in the ways described above. Another type of short note, generally in letter form, gives brief comments on work already published.' A more recent definition specifies that 'they may contain an account of experimental methods and results, sufficiently detailed to allow readers to verify the information ... but with only a very brief introduction and discussion to place the work in scientific context' (ANSI-Z39—16, 1972).

4.4.2 Lists of accessions

The accession lists of libraries (either selective or not) can perhaps be considered as a sort of current awareness. These lists are intended to inform users of the library of new books or journals that may interest them. But accession lists are not selective enough for the special interest of the individual user. Besides which they do not supply enough information, they contain merely the titles of books and journals and not the titles of separate articles, and certainly not abstracts of such articles.

4.4.3 Abstract bulletins or information bulletins

Abstract bulletins or information bulletins go a little further. They do not take the literature of a certain branch of science as a whole but select from the literature for abstracting only those publications that are considered important for a *certain group*.

Selection comes more to the fore here. The differences from accession lists are as follows:
    they contain abstracts, not only titles;
    information about journal articles is included;
    they are based, not on a certain collection in a library but on the world literature or the national literature;
    selection is subjective on the basis of the reader's known or assumed sphere of interest.
Here again, however, the approach is to a group; the specific interests of individual readers are not considered.

4.4.4 Current titles

The editorial preparation of selective abstract bulletins takes much time; there is usually a delay of at least three months after the appearance of the original publications. Lists of current titles appear very quickly, and are published in various forms. The contents pages of a number of journals on a certain subject are collected mechanically and published together in a new journal. The external appearance of such a journal is of no consequence; they are, of course, typographically diverse. There is no attempt at a systematic arrangement. But the reader can get a quick impression (sometimes even before the journals included have been published) of what is or will be published and he can take the necessary measures to procure the material of interest to him on the basis of the article titles.

Selection here is entirely absent, being sacrificed for speed.

4.4.5 KWIC and KWOC indexes

Keyword-in-context (KWIC) indexes are again based on a certain collection of journals or series of reports, and attempt to provide access to the collection by automatic indexing of keywords from the titles of the publications or articles. Their purpose is:
    rapid access to recent literature;
    mechanical processing with as little intellectual work as possible;
    multiple lines of entry to a title within the limitations of the titles.
Here the services have tried to combine speed and access. But here again the approach is to the group and the long lists do not make the tracing of relevant literature easy. The system is dependent, too, on the precision with which authors describe the content of their publications in the titles. Synonyms and foreign languages present further difficulties.

A good example of a KWIC index is *Chemical Titles* published by the Chemical Abstracts Service. The same service publishes *Chemical-Biological Activities* with abstracts.

To overcome objections, some such indexes are made with a predetermined set of descriptors or by adding other descriptors to those from the titles. These may be compiled into KWOC (keyword-out-of-context) indexes. This system requires intellectual preparation by subject specialists.

Sometimes the word KWOC is used in another sense. As in the report 'Supplementary information' to AGRINDEX: 'KWOC stands for Keyword Out of Context, and represents a computer technique by which words are extracted from the English titles of citations and displayed at the head of a listing of citations in which the word appears.'

4.4.6 Selective dissemination of information

Building up the principles of the previous mechanical systems, people have tried to introduce the user's personal interest. Such systems are called Selective Dissemination of Information (SDI). They take 'a profile or list of words describing a technical man's interest'. This profile of interest is expressed in a group of words or combinations of words, and lists of titles or annotated titles are sent weekly or monthly from new material recorded in magnetic form and searched by computer.

Initially such profiles of interest were set down by the users in free language. This however proved inadequate. Firstly there was considerable difficulty with input; secondly the output did not give users what they expected. Users received too many irrelevant titles on their profile. Changes were then introduced in the programs. Users were no longer offered an individual profile of interest but rather had the choice between a number of previously determined profiles. There was thus a switch from pure individual supply to small groups of users who were sent the same material. By this means the costs could be limited.

Since then techniques of machine-processing have been much improved. In general, there has been a return to individual profiles composed for small groups of users. The percentage of relevant titles is now about 70. A difference is indeed made between *Current awareness service* and *SDI*. A service working in a large field of the literature (such as agriculture) that gives rapid information but little selection is called a current awareness service; a service giving more specialized assistance and intended to give a deeper analysis of the relevant literature is called SDI.

Ways of using computers to supply information can still be improved. In the future we can expect more from them. At the moment they are considered as an important aid but—at least in Western Europe—it is believed that an intermediary is necessary between the user and the system. The intermediary (search formulator) must be an expert who can reformulate the users' questions and can critically examine the data churned out by the computer before they are passed to the user.

4.4.7 Alerting service

A particularly individual approach to the user is the alerting service. The success of an alerting service depends on the personal relations between service and user. Before an alerting order can be accepted, an interview is necessary between the user and the service to determine the user's precise needs. Such an interview can be useful only if there is mutual trust and confidence. It must further be possible to alter the order after a period of trial. A second condition is that the service must continually keep the user's interest in view and base its selection from the literature on it. It is even advisable for the literature searcher to visit the user's (research) institute now and then.

A third condition is that the two parties agree on the collection of documents (journals, books, patents) that are examined by the service and that the service checks new material daily and, as far as possible, sends out information daily. A non-mechanical system such as this requires much work and has a limited capacity. It demands further a highly analytical capacity from those searching the literature. In the immediate future, a combination of the intellectual with the computer method is the obvious choice.

4.4.8 Information on subscription

The next step is to set up an information service with a team of information officers, each working for only a few factories or research institutions. It is also becoming increasingly common to have an information officer in a research team. Their job is to keep other members of the team up to date with developments in their field to search the literature. A new term for these information officers is 'the gate-keepers'. Often these information officers act also as secretary of the team and are the editors of their research reports.

In his work, the 'gate-keeper' will use automatic systems (particularly SDI) but he will also need to look up relevant literature.

He must develop a faculty for finding out information and he must be resourceful. As well as studying the literature, he must establish and maintain personal contacts, for instance, by visiting congresses and symposia.

Further reading

Garfield, E. (1972). *Citation Analysis as a Tool in Journal Evaluation.* Science; Washington D.C.

Grogan, D. J. (1973). *Science and Technology: An Introduction to the Literature,* 2nd edn. Clive Bingley; London.

Hills, J. (1972). *A Review of the Literature on Primary Communications in Science and Technology* (Aslib Occasional Publications no. 9) Aslib; London.

Phelps, R. H. & Herlin, J. P. (1966). *Alternatives to the Scientific Periodical.* UNESCO Bulletin for Libraries, **14,** 61–75.

Wood, D. N. (ed.) (1973). *Use of Earth Sciences Literature.* Butterworths; London.

References

American Chemical Society (1961). *Searching the Chemical Literature* (ed. R. F. Gould. A.C.S.). Washington, D.C. (Advances in chemistry series no. 30).

Annual Review of Information Science and Technology, **1–**, 1966–. Wiley; N.Y.

Becker, J. & Hayes, E. M. (1963). *Information, Storage and Retrieval: Tools, Elements, Theories.* Wiley, N.Y.

Bernstein, H. H. & Gabbai, S. (1963). *Inquiry on Non-conventional and Conventional Documentation Systems in Use.* Brussels.

Bottle, R. T. (1969). *Use of Chemical Literature,* 2nd edn. Butterworths; London.

Bottle, R. T. & Wyatt, H. V. (eds) (1971). *The Use of Biological Literature,* 2nd edn. Butterworths; London.

Boyle, P. J. & Buntrock, H. (1973). *Survey of the World Agricultural Documentation Services.* F.A.O.; Rome.

Brewer, J. G. (1973). *The Literature of Geography: A Guide to its Organization and Use.* Bingley; London.

Burman, C. R. (1966). *How to find out in Chemistry,* 2nd edn. Pergamon; Oxford.

Carey, R. J. P. (1966). *Finding and Using Technical Information.* Arnold; London.

CIBA Foundation Annual reviews of communication in science, documentation and automation. Churchill-Livingstone; London.

Crane, E. J. & Patterson, R. M. (1957). *A Guide to the Literature of Chemistry,* 2nd edn. Wiley; N.Y.

Directory of Information Sources in Agriculture and Biology (1971). U.S.D.A.; Beltsville.

Dyson, G. M. (1958). *A Short Guide to Chemical Literature,* 2nd edn. Longman; London.

Fleming, T. P. (1957). *Guide to the Literature of Science.* 2nd edn. Columbia University School of Library; N.Y.

Frank, O. (ed.) (1961). *Modern Documentation and Information Practices.* F.I.D.; The Hague.

Fry, B. M. & Mohrhardt, F. E. (1963). *A Guide to Information Sources in Space and Science Technology.* Interscience; N.Y.

Gates, J. K. (1962). *Guide to the Use of Books and Libraries.* MacGraw-Hill; N.Y.

Gibson, E. B. & Tapia, E. W. (1965). *Guide to Metallurgical Information,* 2nd edn. Special Libraries Association; N.Y.

Grogan, D. (1973). *Science and Technology: An Introduction to the Literature,* 2nd edn. Clive Bingley; London.

Halászi, J. (1968). *The Coverage of Dutch Agricultural Publications in International Abstract Journals.* Pudoc; Wageningen.

Halászi, J. (1968). *The Coverage of Articles from Dutch Biological Journals in Biological Abstracts.* Pudoc; Wageningen.

Herner, S. (1968). *A Guide to Information Tools, Methods and Resources in Science and Engineering.* Herner; Washington.

Johada, G. (1970). *Information Storage and Retrieval Systems for Individual Researchers.* Wiley-Interscience; N.Y.

Jenkins, F. B. (1965). *Science Reference Sources,* 4th edn. Illinois Union Bookstore; Illinois.

Kent, A. (1962). *Textbook on Mechanized Information Retrieval.* Interscience; N.Y.

Loosjes, Th. P. (1973). *On Documentation of Scientific Literature,* 2nd edn. Butterworths; London.

MacKerrow, R. B. (1962). *An Introduction to Bibliography for Literary Students,* 2nd edn. Oxford University Press; Oxford.

Malinowsky, H. R. (1967). *Science and Engineering Reference Sources: A Guide for Students and Librarians.* Libraries Unlimited; Rochester, N.Y.

Melton, M. G. (1965). *Chemical Publications: Their Nature and Use,* 4th edn. MacGraw-Hill; N.Y.

National Federation of Science Abstracting and Indexing Services (1962). *A Guide to World's Abstracting and Indexing Services in Science and Technology.* Washington, D.C.

National Lending Library (1964). *Some Current Review Series.* N.L.L.; Boston Spa, Yorks.

National Lending Library (1966). *KWIC Index to some of the Review Publications in the English Language.* N.L.L.; Boston Spa, Yorks.

Parke, N. G. (1959). *Guide to the Literature of Mathematics and Physics,* 2nd edn. Dover; N.Y.

Pritchard, A. (1969). *A Guide to Computer Literature.* Bingley; London.

Scheele, M. (1967). *Wissenschaftliche Dokumentation; Grundzüge, Probleme.* Schlitz; Hessen.

Smith, R. C. & Painter, R. H. (1966). *Guide to the Literature of the Zoological Sciences,* 7th edn. Burgess; Minneapolis.

Soergel, D. (1971). *Dokumentation und Organisation des Wissens.* Duncker & Humboldt; Berlin.

Troyser, D. L., Kellogg, M. G. & Anderson, H. O. (1972). *Sourcebook for Biological Sciences.* MacMillan, N.Y.

UNESCO (1965). *List of Annual Reviews in Progress in Science and Technology.* UNESCO; Paris.

UNESCO (1968). *Guide for the Preparation of Scientific Papers for Publication.* UNESCO; Paris.

Vickery, B. C. (1958). *Classification and Indexing in Science.* Butterworths; London.

Vickery, B. C. (1965). *On Retrieval System Theory,* 2nd edn. Butterworths; London.

Weiser, S. (1972). *Guide to the Literature of Engineering, Mathematics, and Physical Sciences,* 3rd edn. Donald M. Avery, C.L.R. Distribution Project, Johns Hopkins University Applied Physics Laboratory, Silver Spring, Md. 20910.

Whitford, R. H. (1954). *Physics Literature; a Reference Manual.* Scarecrow; Washington, D.C.

Williams, W. F. (1966). *Principles of Automated Information Retrieval,* 2nd edn. Burns & MacEachern; Elmhurst.

Wood, D. N. (1973). *Use of Earth Sciences Literature.* Butterworths; London.

Yates, B. (1965). *How to find out about Physics.* Pergamon; Oxford.

5 Retrieval of literature

It is reasonable that one can begin a literature search on a given subject in different ways and that the methods of searching out the literature differ according to the purpose of the literature study as already mentioned in Sections 2.2 and 2.4. Nevertheless, we will attempt in this chapter to sketch out a generally applicable scheme of literature searching on a certain subject. We will assume that we have no computer to supply us with information so that we are forced to look up written sources in tracing literature.

The literature search can be divided into three parts:
(*i*) general orientation;
(*ii*) searching on the basis of a systematic plan;
(*iii*) rounding off the search with the most recent literature.
We will consider each of these parts separately.

5.1 General orientation

If we have to seek out literature on a subject that is new to us, the first job is to analyse the problem, to see what the crucial points are and where we can expect to find the most recent literature on the subject. We therefore begin with a general orientation.

For this, we can first look up encyclopaedias and other reference works or indexes. There are specialized encyclopaedias that are likely to be more profitable than the general ones. This orientation will usually yield no more than the titles of a few textbooks and monographs. One must further remember that encyclopaedias and reference works certainly do not yield the most recent literature. One must therefore take note of the year to which the latest reference refers. If we can lay our hands on a guide to the literature of the subject we can make immediate progress. If not, we can go to the subject catalogues of libraries to search for secondary sources. Such a search would cover, for instance, the central library of the university, relevant branch libraries and institute libraries. Subject indexes and catalogues of journals give access to journals which may be of interest.

Initial orientation also includes the tracing of bibliographic sources relevant to the subject. Some of the aids that can be used have been mentioned in Chapter 4. Bibliographies of bibliographies are particularly important. By consulting these sources it is almost certain that we will gain a good impression of how we can set about tracing the literature on the particular subject.

The snowball system

It sometimes happens that one comes upon a recent article on the subject even at this stage. One immediately has an approach to the literature by obtaining the publications in the article's bibliography. On consulting them one can obtain a list of the literature cited in those publications and so continue. In this way one builds, as it were, a tree or network of literature citations. This system is called the snowball system.

This manner of working can lead to rapid results and must therefore not be overlooked. But there are serious objections.

Firstly, one works back through time and never finds more recent literature than the original article. If, for instance, one begins with an article published in 1965, one immediately rejects all literature later than 1965. Actually it is more serious. On average it takes nine months from the submission of a manuscript to the editor of a journal until publication. The chance is also great that the author closed his study of the literature several months before he finally finished his manuscript. It takes time too for the preparation of the manuscript in the organization where the author works: typing of draft, criticism by colleagues or head of department, linguistic correction, drawings of graphs, retyping etc. There is thus easily more than a year's time-lag and literature from 1963 onwards may well be missing.

Secondly, it is all too common for an author to make a biased study of the literature, for instance by taking only national literature or literature in a certain language. If this is so for the original publication seen, the searcher is soon led off the main track. Thirdly, the compilation of a bibliography to an article is to some degree subjective. Some authors knowingly restrict themselves to a minimum taking only articles supporting the evidence or articles that are controversial to their evidence. Other authors fill the bibliography with everything bearing the vaguest relation to the subject. This last type gives especial difficulties, leading the searcher into byways or culs-de-sac.

Citation networks have been used to establish an entire documentation system: *Science Citation Index* which has been published since 1961 by the Institute for Scientific Information in Philadelphia (USA). This system works in the reverse direction from the snowball system. From a basic publication one can trace which publications cite this particular publication. The system therefore works *towards the future*, whereas the snowball system works back *into the past*.

We can now certainly finish the orientation by reading the textbooks and monographs found on this subject and perhaps also some articles, proceedings of congresses or reviews that give us an insight into the problem. The reading and study remain superficial at this stage, and are intended only to gain an impression of the subject and the problems related to it. We need this insight to consider the subject of the search: Is it too broad? Is it too vague? Is it superseded? We end the first phase then by defining the subject of the literature search exactly (that is by writing it down!), perhaps with a short introductory explanation.

5.2 The systematic search

The first phase of searching is thus rather haphazard. Its purpose is to form an idea of the problem, of the amount of literature on the subject, and of how it can be traced. The next phase must be a systematic search according to a defined plan.

5.2.1 The plan of search

From among the bibliographic sources that we have scanned in the orientation phase, we choose those that seem most suitable for our purpose. We must search through these sources systematically over a defined period. It is advisable to discuss the plan of search with the client (or the professor) and then to record the plan on one of the cards of the card system that is to be built up (Section 6.2). In establishing the plan, consideration should also be given to the length of time to be devoted to literature searching and to the writing of the report, and to the date of completion. The plan of search needs to be mentioned in the introduction to any literature survey as well.

How long a period of literature must be searched depends on the subject. If the subject is in a discipline that is developing rapidly there is little point in going back more than perhaps five years. In other disciplines, such as taxonomy, one has to dig deep into history.

Sometimes one discovers remarkable facts. For instance in a literature search on damage by natural gas to trees and shrubs (a modern problem), symptoms were first ascribed to this cause by G. J. Mulder in 1860. It is therefore advisable to discuss with the client or project leader how far back to search the literature.

The systematic search is usually begun with a more complete study of articles in encyclopaedias, by reading parts of textbooks and reference works, monographs and proceedings of congresses. At this stage, the subject catalogues of central libraries, branch libraries and institute libraries can still yield more basic literature.

The next stage is to try (through the bibliographic sources) to lay one's hands on *key articles* or crucial articles. Key articles survey the state of a science, discipline or subject. Such key articles are included in some abstract journals as introductory articles but also occur as separate publications. They often have titles beginning with: *Advances in* ..., (*Annual*) *Review of* ..., *Fortschrifte* ..., *Progress in* ..., *Yearbook of* ..., *State of the Art Report*.... These key articles can often considerably cut down retrieval work. They form a basis on which to continue further systematic searching. For lists of reviews of progress in science and technology see Section 4.2 and references to Chapter 4.

After this, the most time-consuming work usually comes: the tracing of literature through the indexes of selected abstract journals, other current bibliographies and documentation systems in card form. Sometimes abstract journals have cumulative indexes, in other words, indexes compiled from the contents of several years. They give quicker access to information than the annual indexes. The indexes may refer us to abstracts or only to titles of relevant articles. From them, we must make a selection of articles for further study. Selection is easier and more precise if there are abstracts than if there are only titles of the articles.

We can, furthermore, make use of numerous documentation systems in card form. To avoid misunderstanding, let us again stress the difference between documentation systems and catalogues. Catalogues give access to the content of one or more libraries. As a rule they go no further than retrieval of books and journals as a whole. Documentation systems are intended for access to the *content* of journals and books on a certain subject, independently of the place where these books and journals are to be found.

It is necessary here to repeat some remarks made earlier:

Abstract journals have an *average* time-lag of nine months in their abstracting. But sometimes, delays of several years occur after

the date of publication. Title bibliographies are usually faster. Their time-lag is often not more than a month.
(*i*) Abstract journals are forced to be selective. They never cover the literature completely. Coverage is estimated at 60–80 per cent.
(*ii*) Abstracts are made by human abstractors. Subjective influences from the abstractor are therefore unavoidable. It is therefore impossible to be absolutely certain that the abstract covers the contents of the original article.
(*iii*) Abstracts are usually uncritical. No judgement of value can be obtained by reading them.
(*iv*) Documentation systems in card form often give a more haphazard selection from the world literature than do abstract journals. Their composition and arrangement often depend on chance.

It is therefore wise to use different approaches, consulting bibliographies, abstract journals, documentation systems and perhaps catalogues together.

5.2.2 Geography and biography of research

A method that can sometimes be used is the geography of research. By this is understood the following. The more specialized research becomes or the more expensive it becomes (for instance research with nuclear reactors), the fewer the institutions working on the subject. One soon reaches the stage where only a few institutions in the world can specialize in such research.

By consulting annual reports and other publications of these institutions one can find out what is being done in the particular subject. Often the annual reports list the publications of the institutions. By this means, one can also find out where the institution usually publishes so that the tracing of further literature is facilitated. Besides this, one is reasonably sure that one has obtained the good international literature.

One can use the same sort of approach to a subject with only a few *outstanding authors* (biography of research). By consulting author indexes (for instance author catalogues of libraries and author indexes of abstract journals), or *Science Citation Index*, one can check what these authors have published.

5.2.3 Classification principles

While searching for literature one is confronted with different principles of classification or indexing and one soon discovers that there is little uniformity in this field. We will try and give some main principles below.

To oversimplify, one can distinguish two chief means of subject analysis of literature: *classification systems and descriptors*.

Classification systems

In the design of a classification system, one tries to divide the whole of knowledge, a part of human knowledge, or a discipline of science by one or more hierarchical principles.

This designing of a system is called *classification*; classification also covers the placing of documents in an existing classification system (i.e. subject coding, subject analysis or categorization). A (simplified) example of part of a classification system taken from the classification of agriculture in the Universal Decimal Classification (UDC) (634.1/634.8) can make matters clear.

We can hierarchically divide up the term 'Vegetable gardening and fruit growing' as follows:

Vegetable gardening and fruit growing

   Vegetable gardening

  Roots
    Beetroot
    Kohlrabi
    Turnip

    Scorzonera
  Tubers and edible bulbs
    Onion
    Leek

    Shallot
  Plants with edible stalks, leaves or flowers
    Asparagus
    Artichoke

    Brussels sprout

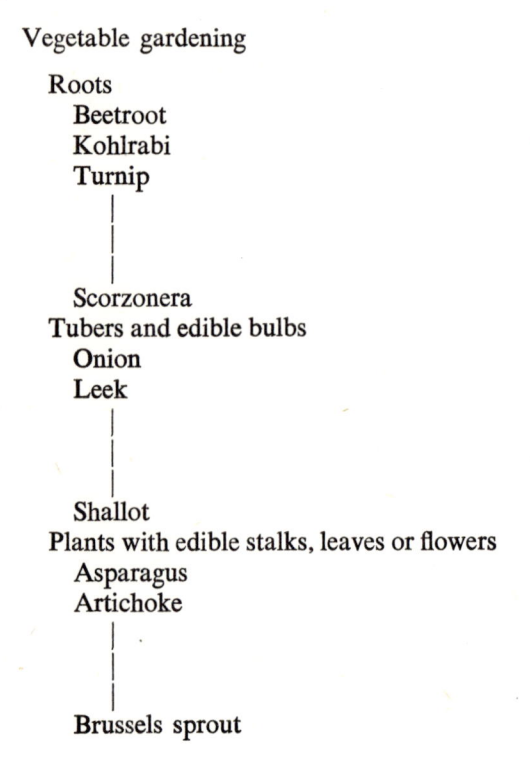

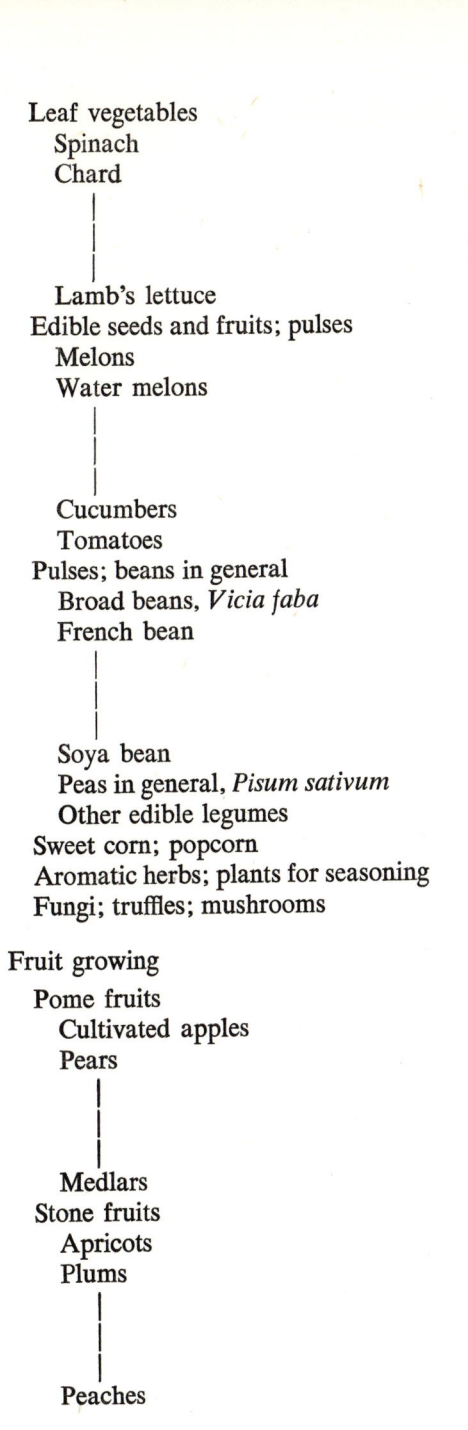

Leaf vegetables
    Spinach
    Chard
    |
    |
    Lamb's lettuce
Edible seeds and fruits; pulses
    Melons
    Water melons
    |
    |
    Cucumbers
    Tomatoes
Pulses; beans in general
    Broad beans, *Vicia faba*
    French bean
    |
    |
    Soya bean
    Peas in general, *Pisum sativum*
    Other edible legumes
Sweet corn; popcorn
Aromatic herbs; plants for seasoning
Fungi; truffles; mushrooms

Fruit growing
    Pome fruits
        Cultivated apples
        Pears
        |
        |
        Medlars
    Stone fruits
        Apricots
        Plums
        |
        |
        Peaches

Citrus fruits, *Rutaceae, Moraceae*
   Oranges
   Mandarins, grapefruit

Moraceae, figs, mulberries
Other pulp fruits
   Annonaceae, soursop, sugar apple
   Myrtaceae, guava, rose apple

Guttiferae, mangosteen
Nuts
   Walnuts
   Chestnut

   Groundnut, peanut (*Arachis*)
Various tropical and subtropical fruits
   Date, coconut
   Olives

   Avocado pear (*Persea*)
Berries and miscellaneous fruits
   Rubus spp.
      Raspberries

      Loganberries
   Ribes spp.
      Red currant

      Gooseberries

    Other shrubs bearing berries
    Strawberries
    Fruits of herbaceous species
        Musa, bananas
        Pineapple, ananas
        Cactaceae, opuntia
        Other fruits of herbaceous species
            Tree tomato
            Passion fruit
Viticulture

On close scrutiny we soon discover that this apparently simple classification contains strange inconsistencies. To take one: a combination of 'tubers and edible bulbs' is perhaps possible to defend (although botanically a tuber is something other than a bulb!) but the combination 'asparagus' and 'brussels sprout' seems strange. However, we cannot easily find another place for asparagus. Furthermore the 'fruits' (in combination with 'pulses'!) eaten as a vegetable (vegetable gardening!) are arbitrarily placed. For 'tomato' this place is indeed acceptable but for 'melon' and 'pumpkin' this section is not above criticism.

If we compare the position in the system of 'bananas' with that of 'strawberries' or of 'oranges' we see that they are not hierarchically equal. We have left out numerous terms about production methods, economics, statistics, machines and equipment. If we include these, the whole system soon becomes very complicated. One must for instance ensure that all sorts of combinations can be built into the system or to take an example, the 'harvesting' (631.55) aspect can be coupled with 'tomato' (635.64:631.55). But in practice such a system is very useful, particularly the Universal Decimal Classification (UDC) as mentioned. The word 'Universal' here refers to the fact that it covers the whole of knowledge. 'Decimal' means that the terms are divided or coded by decimal numbers. The number for 'French beans' for instance is 635.652. Before this one needs to think 0. (nought and decimal point) and so it comes *after* '(0.)635.65 Pulses. Beans in general', and this in turn after '(0.)635.6 Edible seeds and fruits. Pulses'. UDC is an internationally accepted system so that its rules are also internationally determined. It is continually being extended and improved by international committees on special subjects.

There are moderate sized English (BS—1000A) and trilingual (BS—1000B) editions of UDC available from British Standards Institution in London or from Fédération Internationale de

Documentation (FID), the Hague. Detailed schedules on particular disciplines are available from the same sources. There are also many authorized editions available in other languages. Inquiries can be made at the Secretariat of FID, 7 Hofweg, The Hague (Neth).

One should first read through the introduction to UDC and do some 'finger exercises' before using it in practice. At first it seems rather difficult. But one will soon discover that literature retrieval through card systems based on UDC has its attractive aspects.

A commonly stated objection to UDC is that it does not go deep enough for specialists (for instance nematologists). Therefore specialists soon start developing a private classification system, that initially seems easy; but the more the system grows, the more difficulties arise, so that one starts inventing more and more complicated 'house rules'. These house rules make the use of such a private system impossible for outsiders.

In retrieving literature in a documentation system arranged by a classification system it is wise to take the concept somewhat more narrowly and broadly. For instance for 'potato breeding' do not look *only* under 633.491:631.52 (633.491 = potatoes; the colon here indicates the relation between the two terms; 631.52 = plant breeding) but also 631.52 'plant breeding'. There will be publications (for instance reference works) dealing with plant breeding in general with much attention to the breeding of potatoes.

5.2.4 Descriptors

The principles behind the use of descriptors are diametrically opposite to those of hierarchic classifications. The principle of descriptors is that each document (e.g. book, article) can be characterized by (one or more) essential terms called descriptors. An article for instance on 'statistical aspects of selection of cattle' might be characterized by the descriptors: statistics, selection and animal husbandry or cattle. On reading through such an article, one may well add 'breeding value'. These descriptors are then placed alphabetically in an index with a reference to the article after each descriptor, thus in this example four or five times in the index. In a literature search, one would thus find this article under any of the descriptors: statistics, selection, animal husbandry, cattle, breeding value. This method, too, has numerous difficulties.

In a printed index, it is impossible to work with haphazard descriptors chosen separately for every document. This would produce far too extensive and too chaotic an index that could never be consulted. One is thus obliged first to produce a list of descriptors for a certain discipline. It sometimes takes years to produce such a list in a satisfactory form. It is often called a thesaurus (treasury). Guidelines for such thesauri have been established by UNESCO (1970).

Synonyms and homonyms present great difficulty. A choice must be made between synonyms. A second word for any concept may well be included but followed by the reference 'See such and such', a so-called cross-reference. If one, for instance, chooses 'red blood cell' rather than 'erythrocyte', 'red blood cell' is placed in the list of descriptors but 'erythrocyte' is included with the reference 'See red blood cell'.

For homonyms (for instance 'table', which has many meanings), one must indicate the meaning under consideration (for instance 'table (furniture)'). Another situation is if one wishes to refer from one descriptor to another. This can be done with 'See also ...' following the particular descriptor. For instance 'fertilizer, see also manure'. It must be remembered here that by alphabetic arrangement closely related terms are not close together (fertilizer comes under f and manure under m).

In various lists of descriptors or thesauri one goes further and distinguishes between normal descriptors and descriptors with certain connotations:

used for = UF or ;
narrower term than = NTG or <
broader term than = BTG or >
related term = RT or ~
deprecated term = DT.

It is noteworthy that one often sets to work systematically within such lists of descriptors. Descriptors are for instance arranged in categories under headings. There may be subdivision by means of sub-descriptors. For instance in an agricultural index 'potatoes' may form a section taking many pages, the second word being arranged again alphabetically, for instance:

potato, culture
potato, eelworm
potato, management
potato, processing
potato, ridging

This may go as far as descriptors of the third or fourth order so that there is a systematic arrangement on two principles: alphabetic and categoric. In using subject indexes one must take consideration of these difficulties. There is nothing else to do other than to learn the peculiarities of different indexes and systems and to practise their use. The search must be sufficiently thorough to ensure that 'fish do not pass through the net' and one must be sufficiently inventive in trying different approaches.

5.3 Rounding off the literature search

We have only discussed how to find the literature but not how to study it. It is advisable to look for, to consult, and to read selected literature simultaneously. This is partly for a change of activity but partly to obtain a rapid impression of the subject. It also allows occasional use of the snowball system. There comes a moment when one has the impression that no more information can be obtained from the sources used. The literature on the subject seems to be complete, at least the literature back to a predetermined year not too far into the past.

We still have to trace the most recent literature. Here the approach is no longer systematic. Some help is possible with the tools discussed in Section 4.4. We can, for instance, trace recent literature in existing KWIC indexes or by *Chemical Titles*, *Chemical-Biological Activities*, *Current Contents* or similar sources. But ultimately, one can only resort to searching the latest issues of relevant journals that are most likely to contain articles on the subject. In them, one must also look at preliminary notes, letters to the editor and similar items, for instance brief scientific news.

If we find that these latest articles and their bibliographies give no new material we can consider the search closed. The only way of adding new material is to approach specialists on the subject. Such specialists are usually well aware of the most recent advances (through personal contacts). But in this we are moving from the subject of retrospective *literature* information to research information.

Besides personal contacts, we can learn about research work in our own country and other countries by different means. We can, for instance, consult indexes of current research, directories of research institutions in many countries, and annual reports of relevant institutions. All these sources can indicate research going on in a particular country or in a particular subject. One must go to the trouble of writing to the particular institutions or scientists or to visit them to find out what research they are doing and how far it has advanced.

5.4 Summary

In the scheme described, there is the following sequence of operations in tracing literature.

(a) General orientation

(*i*) Consulting (technical) encyclopaedias, reference works, subject indexes and guides to the literature.
(*ii*) Tracing reference books, monographs and proceedings of congresses, in such sources as subject catalogues in libraries. First orientation on the subject.
(*iii*) Tracing the most important bibliographic sources. Tentative view of the amount of literature available.
(*iv*) Tracing the most important journals for study of the subject. Browsing in these journals.
(*v*) Ascertaining the subject of the literature study and fixing the period over which one wishes to study the literature (for instance literature over the past five years). Making a plan of search and a time schedule.

(b) Systematic plan of search

(*i*) A final definition of this subject and of the search plan in consultation with the project supervisor or client. Recording the subject, how far back the literature will be studied and the search plan on the first cards of the filing system and fixation of deadlines.
(*ii*) Final study of reference books, monographs and textbooks. Recording the data.
(*iii*) Tracing and studying key articles. Recording the data.
(*iv*) Tracing literature through indexes of abstract journals and current bibliographies. Selecting from the data found, for instance by reading abstracts. Requesting the primary literature found from libraries. Studying this literature and recording data on cards for one's own card system. (This is the most time-consuming part of the whole operation!)
(*v*) Using the snowball system intermittently.
(*vi*) For completion, consultation of other documentation systems on cards or in other form.
(*vii*) Where applicable, using the geography or biography of research.

(c) Rounding off the literature search

(*i*) Looking out the most recent literature, for instance with KWIC

and KWOC indexes, *Current Contents* and the latest issues of relevant journals.
(*ii*) Perhaps also an approach to specialists in the particular discipline.
(*iii*) Perhaps a check in current research indexes, to see whether there is any research in progress.

Further reading

Garfield, E. (1972). *Citation Analysis as a Tool in Journal Evaluation.* Science; Washington.

References

UNESCO (United Nations Educational, Scientific and Cultural Organization); 1970-07-06. *Guidelines for the Establishment and Development of Monolingual Scientific and Technical Thesauri for Information Retrieval.* Paris (SC-MD—20).
Introductions to abstract journals, classification schemes and bibliographies.

6 Recording information from the literature

A literature report is an unpublished document describing the literature of a given subject and compiled to order. It must satisfy the following requirements.
(*i*) It must include a description of how the literature was collected.
(*ii*) It must describe the literature objectively. Comment may be added to this objective description if the project leader or client specially asks for it.
(*iii*) It must be tailored to the knowledge and interest of the client.
(*iv*) The content must be such that the client can obtain a good impression of the literature treated without reading it himself.
(*v*) The literature report need contain no criticisms of the literature unless the client specially asks for it.
(*vi*) Despite what has been said under (*v*), errors, inappropriate comments or incompleteness in any work must be pointed out in such a way that it is clear that the remark comes from the writer of the report.

6.1 Descriptive and critical reports

One can distinguish two types of literature report: the pure descriptive and the critical literature report ((*v*) above). In the first type, one would usually strive for completeness in the literature of the subject; in the second type one would concentrate on controversial or new ideas in the literature, consider them against one another and draw one's own critical·conclusions. One may even go so far as to suggest topics for further research.

An example of the first type could be a literature report on Farmers' co-operatives in Denmark after World War II. An example of the second type: Comparison between the organization of farmers' co-operatives in Denmark with similar bodies in the United Kingdom.

A report on the first subject could for instance be divided into chapters on: the legal basis of co-operatives in Denmark; the situation of the co-operatives at the end of World War II; the extension of co-operatives to the present day; statistical information on the types of co-operatives, number of members, turnover and finance; the relation between private firms and co-operatives, also with statistical data; the influence of government.

For the second subject, chapters could consider: the differences in legal arrangements; the differences in forms of management; the differences in financing. The advantages and disadvantages could be considered for each chapter and use could be made of published critical articles in the two countries. Finally there could be a chapter summarizing the differences noted and a final conclusion giving recommendations for changes in organization in either the United Kingdom or Denmark.

It should be clear that a different course is set in the former report from the latter, right from the beginning of literature searching, in other words during retrieval, examination and recording of the literature. In the former, it would be important for the reader to find a comprehensive bibliography at the end of the report. In the latter one would have to be selective, collecting only the essential.

6.2 The card system

For each literature report one must build up a card system, in which are recorded data from the retrieved literature. This system must be divided with marker cards. It is thus a systematically arranged card file. The easiest cards to work with are 148 × 210 mm (A5) or 105 × 148 mm (A6).
One sets about the work as follows.

6.2.1 The title card

The title card should contain the following information:
 subject of the report;
 name (and where applicable address) of the client or project leader;
 name of the information officer;
 date of completion of the search;
 classification code or descriptors under which the report can be placed.

Example of a title card
 636.086.74

636:59
619:616-099
Influence of oxalic acid, oxalates and some minerals in leaves and tops of sugar/fodder beets on metabolism of cattle, sheep and pigs.
Client: Instituut voor Rationele Suikerproduktie, Bergen op Zoom, Netherlands
Information officer: T. Eernstman
Date of completion: 1970-12

6.2.2 List of contents

The second card gives the list of contents. This list must correspond with the headings on marker cards.

It is easier to prepare the literature report if the contents arrangement of the report is established as early as possible. While one is working on the system, one will obviously have to amend the subject arrangement. One of the great advantages of a card system over other systems of recording is that changes can be made more easily in a card system by shifting the cards. One must, of course, also amend the card with the list of contents.

6.2.3 Description of the retrieval plan

The third card describes the retrieval plan. The example below, taken from experience in agriculture shows how such a card looks.

Example of retrieval plan card

Subject: Arable farming without tillage or chemical ploughing
Date of completion: 1966-10-26
UDC: 632.954:54 Herbicides (chemicals)
     633.1:632.954 cereals (use of herbicides)
Sources consulted
Subject catalogue, Agricultural University, Wageningen:
  under 632.954:54 and 633.1:632.954. Nothing.
Source file, Pudoc, Wageningen:
  under 632.954. No sources.
5-year card file, Pudoc:
  under 632.954 and 633.1:632.954. Some literature.
Abstract journals:
  Field Crop Abstracts 1963, 1964, 1965 (indexes), 1966 current numbers. Under the descriptors 'Farming systems and practices' (much!), 'Soil and water conservation' (very little), 'Reports' (some), 'Meetings' (some data).

Bibliography of Agriculture 1963, 1964, 1965 and current numbers of 1966. Under 'Soil tillage and tillage' (part of soils and fertilizers). Quite a lot: there is no point in searching further back than 1963.
Departmental and institute libraries:
   Proefstation voor de Akker- en Weidebouw (card system and *collection of reprints*). One or two.
   Instituut voor Landbouwmechanisatie en Rationalisatie. No literature.
Geography of research:
   In Wageningen: Instituut voor Biologisch en Scheikundig Onderzoek, W.A.P. Bakermans.
   In England: ICI Jealott's Hill Research Station, Imperial Chemical Industries Ltd., Agricultural Division, Bracknell, Berkshire. Tele: Bracknell 24701.
   See Field Experiments Guide 1964,
   Plant Protection Guide to the use of Gramoxone W for the renewal and improvement of grassland.
   In France: Licorne, 24 Bd des Italiens, Paris 9e;
   Sopra, 1 Rue Taitbout, Paris 9e,
   See Tuyaux, service agronomique.
Note. ICI, Agr. Division held a symposium on Bipyridylium in Wageningen on 1966-09-28. *Request the report from ICI.*
Sequence of operations
   1. Literature from Field Crop Abstracts (Farming systems and practices) requested.
   2. W.A.P. Bakermans consulted.
   3. Letters to ICI, Licorne and Sopra.

6.2.4 The systematic card system

Then follows the systematically arranged system of cards divided up with marker cards containing the titles of chapters and sections. The cards contain notes from the literature consulted or read. Each card must contain the following bibliographic and other information:
(*i*) name or names of the author(s);
(*ii*) the title of the article, report or book;
(*iii*) description of the periodical or of the publisher;
   ((*i*), (*ii*) and (*iii*) together are called the bibliographic description)
(*iv*) library shelf number (only to avoid a double search);
(*v*) sufficient excerpts from the work that the original publication need not be consulted again during the writing of the report. Exception: an informative abstract is superfluous if one has acquired

the original publication (for instance as a reprint or a photocopy). (Section 6.3 Abstracting)
For bibliographic description of documents (items above), standard rules must be followed. These rules are only nationally standardized. There is considerable work in progress on standard international rules but some proposals have still to be approved (e.g. IFLA, 1971 for books, ISO-R—690 and ISO-R—832). Among the national rules, Anglo-American Cataloguing Rules have wide following in Anglo-Saxon countries. These rules are quite complicated. Usually it is sufficient to give elements in the sequence given below.

Titles of articles in journals

Name of author, initials of author; date (year) of publication. Full title of the article. Name of the periodical, volume number: first and last page numbers. If there are several authors, name also the second and third authors. If there are more than three authors, they may be reduced to *et al.* (*et alii*). The name of the periodical is sometimes abbreviated, but this should be according to standard rules (ISO-4, 1972) or according to a standardized list such as that of *Chemical Abstracts, Biological Abstracts* or *Index Medicus*.

EXAMPLES

Bruinsma, J.; Schuurman, J. J. (1966). The effect of spraying with DNOC (4,6-dinitro-*o*-cresol) on the growth of roots and shoots of winter rye. *Plant and Soil*, 24:309–316.
Stebbins, G. L. *et al.* (1963). Identification of the ancestry of an amphiploid *Viola* with the aid of paper chromatography. *Amer. J. Bot.*, 50:830–839.
Van Luit, B. (1966). Toetsing van koperslakkenbloem en kopersulfaat als kopermeststoffen op bouwland. *Landbouwvoorlichting*, 23:81–84.

Titles of books

Name of author(s), editor(s) or corporate author; year of publication. Title of the book as for journal articles. Edition. Name of publisher (except where identical with the author), place of publication (as stated at the foot of the title page, which should not be confused with the cover!) Number of pages or collation (optional).

EXAMPLES

Influence of man on the hydrological cycle: guidelines to policies for the safe development of land and water resources. = Influence de l'homme sur le cycle hydrologique: aperçu des méthodes utilisées pour la mise en valeur rationelle des ressources en terres et en eaux.

Prepared by the working group on the Influence of Man on the Hydrological Cycle. *Reprinted* from *Status and Trends of Research in Hydrology, 1965-74.* UNESCO; Paris (1972), pp. 31-70, 3½ pp. refs.

Smith, A. U. (1961). *Biological Effects of Freezing and Supercooling.* Arnold; London, 462 pp.

Steen, E. (1966). The effect of fertilizer nitrogen and clover nitrogen on the yield of herbage in Scandinavia. In: van Burg, P. F. J. & Arnold, G. H. (eds.). *Nitrogen and Grassland. Proceedings of the 1st General Meeting of the European Grassland Federation* (Wageningen, 23 June 1965). Pudoc; Wageningen; pp. 77-83; 8 tables; 1 fig.; 23 refs.

De Wit, C. T. (1964). *On Competition*, 2nd edn. Pudoc; Wageningen, 82 pp., 40 figs. (Verslagen Landbouwkundige Onderzoekingen 66.8)

Publisher's and monographic *series* and, for instance, *proceedings* are sometimes treated like journal articles, but they are difficult to trace without details of the publisher, e.g. de Wit and Steen. Full details of a proceedings may alternatively be placed on a separate card with reference under Steen only to 'in: van Burg and Arnold, pp. 77-83'.

If there is no named personal author or editor, take the name of the corporate body responsible for the work or the first word of the title (ignoring or transposing articles and numbers), for example 'The First International Congress on Space Travel' is placed under: International, Congress on Space Travel, 1st.

For books and series, the bibliographic description can be copied word for word from the catalogue cards of the library. Although not strictly essential, a bibliographic description can include collation, e.g. number of figures (fig.), tables (tab.) or references (ref.); also useful to know is whether it is a review article (rev.) or a translation (transl.). Many recent books have a standard number (ISBN) and this should be mentioned after the collation.

6.2.5 The author index

Although it would seem to demand extra work, it is useful to repeat the bibliographic description and to place the second card alphabetically under the (first) author in a separate card file. From this the list of references can easily be compiled, being arranged alphabetically by author.

6.3 Abstracting

To make a record of the content of the literature, one must abstract it. An abstract is a short objective reproduction of the content of a document (DIS—214; ANSI-Z39—6).

There are, unfortunately, different words that are used for the concept of an abstract, in which people try to insert different nuances. We will not go into these nuances but will regard the words summary, synopsis, resumé, excerpt and abstract as synonyms.

The terms 'author's summary' or 'author abstract' which refer to an abstract written by the author himself are worth separate mention. Some doubt the value of such abstracts; others are proponents of them. Users of the literature would certainly be wise to consider an author's abstract more critically than one made by a third party. If it is taken over, it should be marked 'author's abstr.'.

6.3.1 Indicative and informative abstracts

A distinction is made between indicative and informative abstracts. An *indicative* abstract is one summing up the content of the publication. Such an abstract answers the question 'What is it about?' or, at most, 'What has been done?'. An *informative* abstract must contain the actual facts or thoughts, perhaps also methods of research, results, conclusions or recommendations and thus answers the question 'What was discovered?' or 'What was achieved?' or 'What must be done?'.

If one only gives a short description of the content under bibliographic description, for example one or two lines, this can be called an annotation, as already mentioned. The length of the abstract is not of vital importance. Services publishing abstract journals usually make longer abstracts of articles that are difficult to obtain or read (for instance Japanese articles) than of articles that are easily accessible. The length of the abstract is thus no indication of the scientific value of the article. But in practice, there are some limitations. It is preferable for instance to make a text that can easily be put onto an A7 card (74 mm × 105 mm) or an International Library Card (75 mm × 125 mm). For machine processing by punch card, for instance, one is again restricted to a certain length by the number of punches that are possible. With modern techniques, this last limitation is less essential.

If one is making abstracts for one's own use, length is of less consequence. It may even be desirable to include tables and figures in an informative abstract, and in an indicative abstract, to list the titles of some chapters or sections. It is worth remembering that modern methods of reproduction allow us to copy tables, figures, lists or whole chapters of text with little effort.

If abstracting for someone else (for instance for an abstract journal), one should generally try to give a *complete* picture of the content of the publication; when abstracting for oneself one can work selectively and only take over such material as is of interest.

In both cases one must clearly distinguish what is taken over from what one adds, for instance criticism. In published abstracts it is usual to place criticism or comments by the abstractor between [square] brackets and signed 'Abstr.'.

6.3.2 Abstracts for one's own use

Before making an abstract for one's own use, one must decide if it is to be indicative or informative. Remember that the abstract is an aid to memory. An *indicative* abstract suffices if the original article is readily available (for instance because one has made a photocopy) and also if the article is only to be mentioned in passing in the literature report. An *informative* abstract is necessary if one cannot or does not want to refer back to the original. If so, the abstract must contain everything needed later. If there are notable quotations one is wise to copy them in full and enclose them in quotation marks.

To make an (informative) abstract of a primary article for one's own use one sets to work as follows:

(*i*) Read the title and then the summary and judge whether these contain sufficient interesting aspects.

(*ii*) If so, look for the scope note (or statement of purpose in the introduction) and the conclusions and read them.

(*iii*) If the article still seems of value, read over the whole article. Try to 'read vertically' and to digest descriptors or index words. Sometimes one can pass over whole sections (for instance methods of research or part of the factual material that is of no interest to you for the report).

(*iv*) After reading over the article, read selectively and as far as possible make notes simultaneously for the abstract.

(*v*) Then write the abstract with an intentional sequence (not necessarily the sequence of the article). It can for instance be practical to use a sequence answering the following questions: why—what—how—what results?

EXPLANATION

why: answer to the question why the research was begun
what: scope and objects of research, possibly with a statement of where the research was carried out
how: experimental method
what results: the results, discussion of results and conclusions.

This arrangement corresponds closely with that of Rothkirch-Trach in his 'Positionsreferat'. The principle of his argument is as follows. The 'modus procedendi' in science follows a chain of ideas. This chain can be thought of in five links (categories or facets), which he names 'Positionen' (positions). He defines these as follows:

1st Position: scientific discipline.

> Every experiment fits into a discipline or combination of disciplines. It is therefore desirable to include the discipline in the abstract.

2nd Position: scope or purpose.

> Every experiment and every scientific review has a question as basis. This question of theme must be mentioned.

3rd Position: the actual object of research.

> In the natural sciences, one works with actual objects, which must be described. In the social sciences and humanities, this is less common.

4th Position: experimental methods.

> In general, nature does not give a straight answer to the question. Experiments are necessary by certain methods, which must be described in order to know how one arrives at the results.

5th Position: results.

> Every experiment gives either positive or negative results. These results can be expressed in numbers, formulae or concepts (words). These results are often the most important part of the publication. It is of course not always necessary to include all these elements; this scheme is more a guideline along which one can work. On the other hand, one sometimes needs to record more about certain positions than can be included in an abstract. If so, one can photocopy large parts of the document, number these photocopies and store them by number. One must then give the reference numbers in the abstract.

6.3.3 Language and style of abstracts

In general, abstracts in telegram style are viewed with disfavour. Apart from their poorer readability, this style can lead to ambiguity and imprecision.

A few rules worth observing in abstracting are:
not to repeat information given in the title;
to use a direct concise style. Not: 'For uncondemned raw materials that reach the disposal unit, it can be stated that there is only a complete guarantee if human health is not brought into danger', but: 'There is a complete guarantee for uncondemned raw materials that reach the disposable unit only if they do not endanger human health';
to avoid passive and oblique forms of verbs. Not: 'It was found that there was no relation', but: 'There was no relation';
to avoid expressions that say little, such as 'in this article' or 'some aspects of';
to avoid unexplained abbreviations;
to put quotations from the text between (inverted) commas.

6.3.4 Examples

Below are various types of abstracts made from one article.

Annotation
Kreyger, J. (1966). Behoud van caroteen in kunstmatig gedroogde groenvoeders (Preservation of carotene in artificially dried forages). *Landbouwkundig Tijdschrift*, 78: 33–37.
   Discussion of factors influencing the carotene content of dried forages and of ways of limiting the decline in content.

Indicative abstract
Kreyger, J. (1966). Behoud van caroteen in kunstmatig gedroogde groenvoeders (Preservation of carotene in artificially dried forages). *Landbouwkundig Tijdschrift*, 78: 33–37.
   From differences between contract driers and commercial driers the author introduces differences in the products produced by the two groups. After discussing the theoretical principles governing the manner and amount of carotene broken down he considers measures that can be taken to preserve carotene during storage.

Informative abstract
Kreyger, J. (1966). Behoud van caroteen in kunstmatig gedroogde groenvoeders (Preservation of carotene in artificially dried forages). *Landbouwkundig Tijdschrift*, 78: 33–37.
   In the Netherlands there are two categories of drier: co-operative contract driers and co-operative or private commercial driers. Contract driers try to produce a high content of protein as well as

a high content of carotene; commercial driers aim particularly for a high content of carotene. Both commercial and contract driers now supply a product that reaches reasonable standards for protein digestibility and for carotene content.

Carotene is oxidized enzymically, photochemically and thermically. The first two processes occur between mowing and drying; the last during storage of the dried forage. Thermic breakdown proceeds more slowly in forages than in pure carotene, perhaps because of naturally occurring anti-oxidants. The rapidity of breakdown of carotene is directly related to the content at that moment. The rate of breakdown roughly doubles for every 10°C increase in temperature of storage. Means of conserving carotene during storage are:

cooling before and during storage;

use of anti-oxidants—a suitable one is ethoxyquin (trade name Santoquin, Monsanto);

storage under inert gas. Until recently, this method had not been used in the Netherlands.

Informative abstract for one's own use

On the specific subject: breakdown and conservation of carotene.
Kreyger, J. (1966). Behoud van caroteen in kunstmatig gedroogde groenvoeders (Preservation of carotene in artificially dried forages). *Landbouwkundig Tijdschrift*, 78: 33–37.

Breakdown of carotene is by oxidation. There are three processes:
(*i*) Enzymic breakdown. This reaction is rapid particularly with severe crushing. The enzymes occur in green leaves.
(*ii*) Photochemical breakdown. This is less rapid.
(*iii*) Thermic breakdown. This occurs in dry leaves; it is slower than (*i*) and (*ii*).
[Further details of rates are lacking. Abstr.]
Processes (*i*) and (*ii*) are important between mowing and drying; process (*iii*) in the storage of dried forage.
Thermic breakdown in the dry leaf is slower than carotene as a pure substance. See Refs 7, 8, 9 (Photocopy 16).
There are two governing factors: (*a*) rate of breakdown is directly related to content; (*b*) rate of breakdown roughly doubles for every 10°C rise in temperature of storage. Because of Factor (*a*), the concept of half-life has been introduced.
For a half-life, temperature must be stated. See Table 3 (Photocopy 17).
Carotene can be conserved during storage by the following means:
(*i*) Cooling before and during storage. A storage temperature as low as possible is particularly important. For this the product must be

well cooled after drying and, during processing, night air can be introduced into the storage place. This should only be done if the outside air is at least 6 or 8°C lower than the temperature of the product (danger of too wet a product).

(*ii*) Use of anti-oxidants. Suitable is ethoxyquin (2,2,4-trimethyl-6-ethoxy-1,2-dihydroxyquinoline; sold as Santoquin, by Monsanto) applied to supply 0.15 g a.i. per kg dry product. For this, dilute the ethoxyquin in emulsifiable form with 50 parts of water (ca 10 g/litre) and apply 100 litres per ton by pumping it over the forage before milling. The rate of breakdown is roughly halved. Costs up to Dfl. 0.25 per 100 kg dry product.

(*iii*) Storage under inert gas. This method is used in the United States, but not in the Netherlands (too dear). In France, there is one large central storage installation with inert gas. It uses generators which take oxygen out of the air by burning of oil or gas. The residual air with the oxidation gases forms an inert mixture.

For storage the product must be at ambient temperature; moisture fraction in the pellets may not be higher than 0.06–0.08.

Further reading

Standards mentioned are listed in Section 7.11

Maizell, R. E., Smith, J. E. & Singer, T. E. R. (1971). *Abstracting Scientific and Technical Literature:* An Introductory Guide and Text for Scientists, Abstractors and Management. Wiley; N.Y.

Rothkirch-Trach, K. C. von (1966). Methodische Textanalyse—Referiertechnik im Zeichen der automatischen Dokumentation. *Nachrichten für Dokumentation*, **17,** 169–71.

Weil, B. H. (1970). Standards for writing abstracts. *Journal of the American Society for Information Science*, **21,** 351–57.

7 Writing the literature survey

In this chapter we will describe how to write the *literature* survey as an example of how to write reports in general. Specific requirements of research reports and of the writing of advisory articles about results of research will be dealt with in later chapters.

7.1 Different procedures for writing

It is incorrect to say that there is only one method of writing a good report. But it should be said that, whatever the method of writing the report, the result of that writing in the form of a report must satisfy certain requirements. One writer will however prefer one way of approaching this while another will prefer some other way. Below are some examples.

7.1.1 The impulsive method

Some writers can only formulate their ideas by putting them on paper. This is, so to speak, thinking while writing. Writing is spontaneous and rapid. It does not consider arrangement and faults of style or language; it follows grasshopper logic, the elaboration of ideas coming suddenly into the writer's head, even though they are incoherent; and finally a full stop is added at the end of the whole work. The real work begins later: the arrangement and rearrangement, the pure formulation and knocking the report into shape. Clearly the first version must be entirely rewritten at least once and preferably more than once; this method does not, therefore, save time. Its advantage is its spontaneity, and often its personal character and the worth that one finds in it. The great disadvantage is that the writer often remains bound to his original version so that the final report contains traces of immaturity.

7.1.2 The pieces method

Many writers have the greatest difficulty with the start of the report, the introduction. This difficulty can be circumvented by not starting at the beginning of the report but with various easily surveyable pieces such as the paragraph concerning the historical development of the subject. Of course with this method one must know roughly beforehand what the shape of the report is to be. When one has worked out all the pieces in this way there comes the work of putting it together and making a start and finish to it. The thinking is largely complete and experience shows that one can then easily formulate the introduction and conclusion.

The advantage of this method is that one is working with pieces that can easily be comprehended. One disadvantage is that one may doubt where a certain aspect should be placed; another is that the logical ordering can be lost, and that there is difficulty in making the report into a coherent whole. Partial rewriting will often be necessary.

7.1.3 The method of the inverted pyramid

To elaborate a strictly logical account it is advisable to begin at the end: the formulation of conclusions (by points!) at least in tentative form. One begins with the top of the pyramid. One then has a fixed aim in view and it is no longer so difficult to hold on course. It is then not important whether the final report places conclusions first or at the end. One must, however, do the real thinking beforehand but that brings benefit to the real writing.

The advantage is that writing proceeds more easily, but the great disadvantage is that one must first draw conclusions from a large unsystematic collection of information. For many this is a difficult obstacle.

7.1.4 The systematic method

In the systematic method one elaborates the report in stages. Starting with the basic material one attempts to reach an arrangement by a system of descriptors. One only starts to write when this is defined. All that one then has to do is to clothe the skeleton. The advantage is that the report is easy to write and there is reasonable chance of achieving a well balanced report. The disadvantage is that one can be confined by the system at the expense of the spontaneous personal character of the report.

Since this systematic method offers the best chance of a good report for most authors, we will further elaborate it. When the data have been collected from the literature and filed in the system, for instance, in the form of abstracts, quotations or notes, this is already a good basis for the system. The file can then be divided up by tabcards and the cards can be easily arranged in the order in which the data are to be dealt with. The first step is to characterize each item by a descriptor (which may consist of more than one word) and to note these descriptors in a rough draft (if desired, on cards). Further descriptors are added for thoughts and facts that one wishes to put into the report. The more complete this list the better. At a later stage, some of them may well be scrapped.

The same procedure can be used for other types of reports. There too, one looks through all the data that has been collected and, for each group of data or each idea, one notes a descriptor. In both cases one runs through these descriptors several times and tries to group them in what will later become the sections.

The next step is to think up a heading for each group of descriptors and place these headings in a logical sequence, so determining the tentative divisions of the report. Having progressed so far, the descriptors must be placed under each heading likewise in a logical sequence so that the scheme of the report is fixed.

The next step is important: deciding on the size of the report. The size must be fixed *before* one starts writing. The most important consideration in this is the question: What is a reasonable number of pages that the client (or the would-be reader) can read through with interest? A second consideration must indeed be: What is the minimum number of pages necessary to deal with the subject purposefully? The size of the report can best be expressed in the total number of words. This is not difficult if one has decided on the number of pages. Even in handwritten text, the number of words per line is almost constant; it may, for instance, vary between 12 and 14. It is sufficient to determine an average for 10 lines of written or typed text. Typed text can easily be converted to printed text, for instance, in the same way (as long as the method of printing and the format of the printed work are known).

Once the number of words is known, this can be divided among the paragraphs or sections, taking note of the importance of each section in the whole report. If it is desirable to go even further into it, one can decide how much must be devoted to each descriptor within the paragraph. Only when matters have progressed so far does one begin with the writing, taking careful note of the scheme including the number of words. In practice, little need be done to polish the report afterwards.

An elaborated example of this procedure is included in Section 7.12. The example is based on the preparation of *this* chapter.

7.2 Principles of text arrangement

In the arrangement of a report into chapters and sections, one must proceed from previously selected principles and take care to stick to this principle. After deciding on, for instance, chronological arrangement, one must not in a later section of the report introduce a systematic treatment. Some examples are given below to give an impression of principles that can be used.

The theoretical principle

The literature survey is built on a theory. In other words, one begins with a general theoretical consideration and considers parts of it in detail one by one. In each part, opinions for and against the theory are given where possible with one's own judgement. At the end of each consideration, one gives with figures the evidence for the theory.

The logical principle

This principle is used in almost all research reports and can also be used in certain types of literature survey. It is as though a straight line runs through the report. It begins with a description of the problem and then follows strict proof based on available data leading step by step to the conclusions.

The chronological principle

The report begins with the description of the earliest literature consulted and proceeds through the whole literature up to the most recent.

The didactic principle

The report proceeds from the most readily envisaged: the application and proceeds to theoretical considerations and to the more abstract fundamental basis.

The psychological principle

The report begins with what is essential for *the user* and afterwards deals with details and theory.

7.3 Some basic rules

For a critical assessment of one's own report, the following rules can be used.

Cross-references

A poorly arranged report can often be recognized by numerous cross-references. One often finds phrases such as: 'As will be further described in Section ...' and 'As already elaborated on page ...'. This is of course different from the situation when the introduction describes how the report is arranged and where different things can

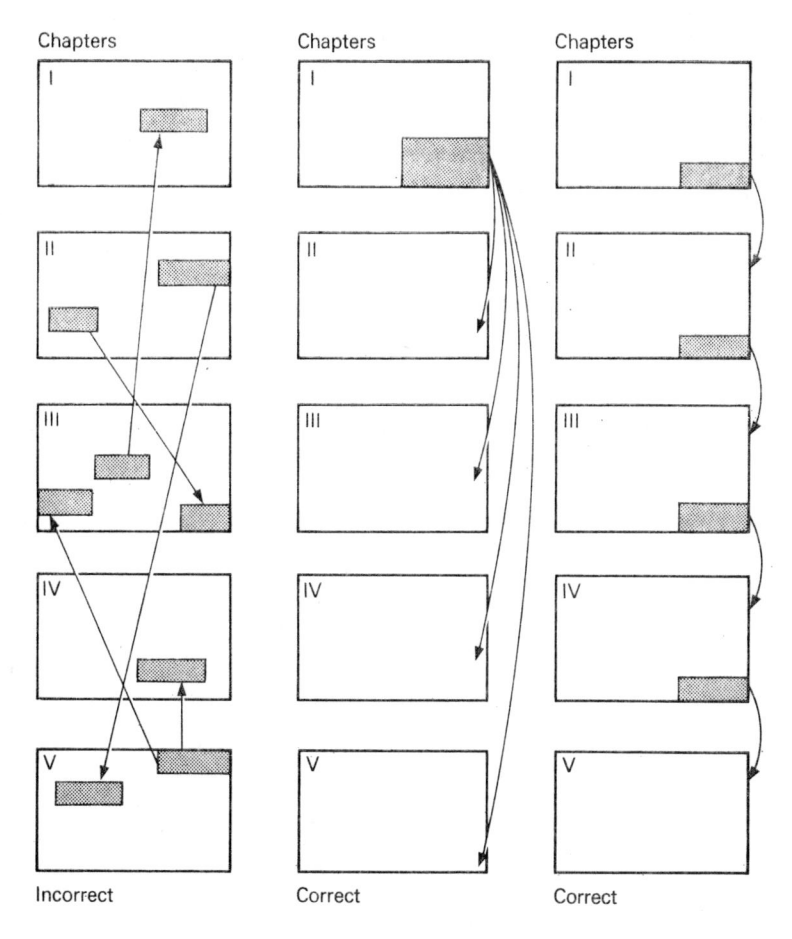

Fig. 6. Ways of cross-referencing in a report.

be found in the report. Another reasonable system of cross-reference is when one refers on from the end of one section to the next section (see Fig. 6).

Conclusions before evidence

One must make sure that conclusions are not drawn before the evidence has been given. An exception to this rule is, of course, if conclusions are deliberately and consistently given first and then the evidence (as for instance in textbooks). But then a clear division must be made, so that no doubt is possible.

Elaboration of pros and cons

A literature survey is almost unreadable if all the literature is enumerated in turn. Attempt to weigh up the opinions of different authors against one another. This can be done in each small section by weighing the opinion of A against P; one can also give the opinion of A, B, C, D, etc. and then the divergent or contrary opinion of P, Q, R, S, etc. The preference depends on the situation. Here are two comments:

(*i*) The reader expects the writer of the report to take a stand, or at least to offer an opinion. It is particularly annoying for the reader of the whole report to have a continuous pronouncement of 'It might be frosty; it might be mild'.

(*ii*) A clear distinction is necessary between what is taken from the literature and one's own arguments. There must be no possible doubt. This applies equally in the description of one's own research.

Selection from the literature

The writer of a literature survey can easily be tempted to pack as much as possible of the literature he has traced into the report. Indeed, the writer has taken the trouble of looking up, studying and abstracting all this literature, and now he wants to show the reader how much work he has done. But a report gains in authority if one is selective and sticks carefully to the main theme. Sticking to the main theme becomes easier the more the purpose of the report is defined beforehand. Vague definitions such as 'Some factors governing the relation between yield and quality' or too broad a subject such as 'Pollution of surface water' cause one to run off the main theme and lead to inconsequential reports.

Closing off the report

A report should not end in the air; in other words it should not end, for instance with the last detail of the last citation. Take care to bring the report to a close, if possible with conclusions or hypotheses. If the literature can offer no solution to the problem that had been chosen as subject, one may even give the state of the art when closing off the report. One may possibly suggest ways in which a solution could be found.

Explanation

A literature survey should begin with a short section on how the literature search has been carried out. The card with the search plan (Section 6.2.3) forms the basis for this paragraph. The paragraph is usually entitled 'explanation'. This section is important for the client because he can then see what literature has been covered and what literature has not. But, for the writer of the report too, this section is important. Should he later do a related or supplementary search, he should be in no doubt as to what he has already done.

EXAMPLE

The 'explanation' for a literature survey on 'Arable farming without tillage' for which the plan of search was given as an example in Section 6.2.3 could be as follows:
'The indexes of *Field Crop Abstracts* were searched for the volumes 1963 to 1965 and current numbers of the 1966 volume. There proved to be a considerable amount of literature, particularly under the descriptor "Farming systems and practices". Quite a lot of literature was traced too through *Bibliography of Agriculture*, for which the volumes of 1963 to 1965 and current numbers of 1966 were searched for the descriptor "Soil tillage and tillage".

The systematic catalogues of the Landbouwhogeschool Library Wageningen, the source file and the quinquennial card file at Pudoc produced only a few supplementary publications. By contrast, a very important personal contact was Dr Ir W. A. P. Bakermans who is investigating arable farming without tillage. He told us among other things of a symposium held by the Agricultural Division of ICI at Wageningen on 28 September 1966 on Bipyridylium and gave us the addresses of Licorne and Sopra, two organizations in Paris that are concerned with this problem.

Letters to ICI, Licorne and Sopra produced important documentation material which has been included in the following report.'

Synopsis

It is moreover desirable to accompany a report by a synopsis which is something other than the concluding section! The synopsis must give a very brief survey of the *whole* report. It can be placed at the back or front of the report. It is becoming more and more usual to place the synopsis at the front on a separate page, directly after the title page.

Letter of transmittal

The report must be transmitted in some way or other to the client. Of course one can often hand over a report without further ado but it is still advisable to accompany the report with an informative letter. Firstly this offers the opportunity of determining the date of transmittal; secondly the letter offers the chance of putting various points into writing.

The letter of transmittal should be written in the 'I–you' form. As a rule it defines the project the writer has been given, where necessary by reference to discussions that took place during the compilation of the report, to a few important results from the literature survey and perhaps to a few recommendations for further research.

7.4 Subheadings and section numbering

According to the length of the report it can be divided into chapters, sections and subsections or only into sections and subsections.

Typography makes the necessary clear distinctions in the hierarchy of headings. One can distinguish headings, for instance, by choice of letter (capital or small), by the position of the heading in the page (centred, flush to the left and on a separate line or in the line) (displayed or undisplayed) and by means of space (lines left blank) between heading and text. Restrict underlining to a minimum as a means of distinction.

Arrangements can be made with Roman numerals, Arabic numerals and letters, but a decimal arrangement as recommended in ISO-R—2145 is also good. For the first system, use large Roman numerals for chapters, Arabic numerals for sections, and small letters for subsections, thus I, 1 and a. For the second system one has: 1, 1.1 and 1.1.1. With the decimal system, one should certainly not go further than 1.1.1.1.

Headings and text should be readable separately from one another. One should not, for instance, begin as follows:
(Heading) '*Effect of improving varieties*'
      'This shows that productivity per year ...', but one should write:

(Heading) *'Effect of improving varieties'*
  'The effect of varietal improvement shows that productivity per year....'

Further, one should try to keep the headings in a uniform style. Not, for instance, as headings: 'Increasing productivity', 'The improvement of varieties', 'What has been achieved since 1955', 'The self-pollinators and the cross-pollinators', but: 'Raising productivity', 'Varietal improvement', 'Progress since 1955', 'Self-pollinators and cross-pollinators'.

7.5 Use of language

Let us assume that the literature survey is written in English. For those using English as a second language there are special problems. One must first learn to write well in one's own language before one can write well in a second language. There are innumerable books on the writing of English. There are also many manuals and guidelines on report writing, many of which are of American origin.

Council of Biology Editors, Committee on form and style (1972). *Style Manual*, 3rd edn. American Institute of Biological Sciences; Washington.

Fowler, H. W. (1965). *A Dictionary of Modern English Usage*, 2nd edn revised by Sir Ernest Gowers. Oxford University Press; London.

Gowers, Sir Ernest (1973). *The Complete Plain Words*, Rev. edn. HMSO; London.

Roget, Peter Mark (1852). *Thesaurus of English Words and Phrases*. Available in many editions, for instance by Penguin, Harmondsworth, Middlesex.

For British spelling the usual authority is:

Fowler, H. W. & Fowler, F. (1964). *The Concise Oxford Dictionary of Current English*, 5th edn revised by E. McIntosh; etymologies revised by G. W. S. Friedrichsen. Oxford University Press; London.

The Shorter Oxford Dictionary and the full *Oxford English Dictionary* are less used as standards for spelling.

Of the American dictionaries Webster's *Third New International Dictionary* is widely accepted, containing also useful information on synonyms and near synonyms, and introductory articles on spelling and punctuation.

Also very good are:

Morris, William (ed.) (1969). *The American Heritage Dictionary of the English Language.* American Heritage Publishing Co. Inc.; New York.
(This dictionary contains useful notes on acceptable usage.)

Treble, H. A. & Vallence, G. H. (1936). *An ABC of English Usage.* Oxford University Press; London.
(This book is simpler, cheaper, and less dogmatic than Fowler.)

Collins, F. H. (1956). *Authors' and Printers' Dictionary.* Oxford University Press; London.
(Revised 1973).

Partridge, Eric (1970). *Usage and Abusage: a Guide to Good English.* Penguin; Harmondsworth, Middlesex.
(A useful and shorter guide than Fowler.)

There are notes on punctuation in Fowler, Treble & Vallence and Partridge.

A useful complete work on the subject is:

Carey, G. V. (1971). *Mind the Stop*: A Brief Guide to Punctuation with a Note on Proof Correction. Cambridge University Press; London. Reissued by Penguin; Harmondsworth, Middlesex.

Woodford, F. P. (ed.) (1968). *Scientific Writing for Graduate Students.* Rockefeller University Press; New York.
(Though intended primarily for teachers of scientific writing, this book is also extremely useful to students.)

When reading different reports or articles, it is striking that one is much easier or pleasanter to read than another; the one publication is more *readable* than the other.

The American Heritage Dictionary defines readable as:
   capable of being read easily; legible;
   pleasurable or interesting to read.

We can thus ascribe three meanings to readable:

(*i*) Something is readable when the writing is clear (for instance a readable manuscript); this can also be called legible;

(*ii*) Something is readable when the content of the writing is intelligible for the reader (for instance this article on nuclear physics is readable); this can also be called reader interest;

(*iii*) Something is readable if language and style are so chosen that it reads easily (for instance this writer has a readable style); this can also be called reading ease.

These three meanings are related. A well written and clear publication can be unreadable when it is, for instance, printed with too small a letter or on poor paper (for instance some cheap paperbacks!). A typographically well produced book can be unreadable for a certain group of readers if the subject is unintelligible to them (for instance, a scientific book on genetics). A work on a simple topic can be unreadable because of complicated grammatical constructions.

Below we restrict ourselves to the third meaning of readable, in other words easy to read. As an explanation, let us take a (translated) example from Landbouwkundig Tijdschrift:

'If on the one hand we extrapolate the line of increase in wages—as it was sketched by Mr Hupkes—and on the other hand we see the development of combine harvesters with a grain tank and of completely automatic potato and beet harvesters that put the product direct into the truck, then it is clear that the amount of work on our arable farms has decreased at a very rapid rate and will continue to decrease even further.'

This sentence contains 79 words and is complex. It contains the constructions 'on the one hand ... on the other hand' and 'if ... then'. Besides that, it contains two interwoven ideas: increase in wages—mechanization—decrease in work; and harvesters—further development with tanks and complete automation—delivery of product into truck.

The ideas in the sentence can be expressed as follows:

'Hupkes has sketched the line of increase in wages. We may extrapolate that line. We see too that harvesters have developed further. Combine harvesters now have a grain tank; automatic potato and beet harvesters deliver the product direct into the truck. The increase in wages and the improvement in harvesters have rapidly reduced the amount of work on our arable farms. The prospect is that this reduction will continue.'

The long sentence has been broken into six sentences of $9-5-(7+12)-20-8$ words and is thereby easier to digest, even though there is little reduction in the number of words (69). The difficulty remains that the two ideas are placed in such proximity that a jump in thought is necessary.

Another example from an article:

'If the size of the holding is taken as a fixed item of data, it can be considered whether to achieve the same aim by irrigation and by feeding in stall.' The sentence can be rewritten as follows:

'If we take the existing size of holdings as a fixed item of data, we can consider whether to achieve the same aim by irrigation and by feeding in stall.' Turning the sentence from passive into active mode makes it read more easily. The number of words is unchanged.

An artificial example explains the difficulties with long 'classical' words and loose linkages. 'Incomparable bibliographical measures were undertaken for the exhaustive and critical elimination of unutilizable cases of apparently uninspired and unharmonious tautological paraphrases and circumlocution, predisposed by somewhat moribund syntactical pedagogical facilities, by interlocutary autographical perliniarization, complemented by re-orientation of ambiguous organization, clarification of concepts, demythologization, comprehensive simplification of neologisms and determinologization.'

The reader is held up by the long words, the unclear relationship between the various parts of the sentence, and by the mass of abstract nouns. The function of science is the collection of all facts on a subject and the finding of the briefest way of clearly describing the facts. Though readability may be increased by brevity, it is certainly not increased by a telegraphic style. It may be helped by splitting long sentences into short sentences or by replacing long words of Latin or Greek origin by short native words. It may be helped by use of active verbs. 'We studied something' is far easier to read than 'experiments were carried out on something'. Particularly important are brief clear links between the parts of a sentence in the form of prepositions (of, by, from, to) and conjunctions (and, but, however, although, which, that, because).

Hundreds more examples could be given. But the point to notice is which factors need attention in order to avoid unreadable text. Below we shall name a few.

7.6 Factors in readability

7.6.1 Length of sentences

As sentences become longer, they usually become more complicated and more difficult to read. Long sentences are almost bound to follow if one begins with 'Although ...' or with constructions like 'on the one hand ... on the other hand', often with many clauses and phrases. One can also go to the other extreme with too many successive short sentences. This is particularly irritating if all the sentences have the same construction: subject—verb—predicate. For the first example, we could, for instance, write:

'Hupkes sketched the trend of increase in wages. We have extrapolated this trend. Harvesters are developing in the direction of combines with a grain tank. Potato and beet harvesters deliver the product directly into the truck. The amount of work on arable farms is rapidly decreasing. This amount of work will decrease still further.'

The change gives very short sentences but is no more pleasant to read. The best way is to intersperse long sentences with short sentences and to vary the construction of sentences.

7.6.2 Choice of words and terms

The unfamiliar word is often an obstacle to smooth reading and understanding of the text for the general reader. If one has a choice between an unfamiliar word and an everyday word, choose the everyday word, for instance red (blood) cell instead of erythrocyte or gut(s) instead of gastro-intestinal system. Do not worry too much about subtle differences in meaning.

Scientists tend to hold other opinions. It is suggested that internationality of science is associated with internationality of language and that it is twice as much effort to put the unfamiliar word into native language and then to translate it into a foreign language. There is something to be said for this view with technical terms as long as the technical terms are restricted to the international group of specialists.

But if our job is to translate scientific results into recommendations for farmers or for industry, the unfamiliar word can often be an obstacle. Often the spelling and pronunciation of strange words in itself is a difficulty between people of different nationalities. In any case, scientific writing consists largely of ordinary words and this opinion is therefore not valid. So long as we can speak of a native language it is advisable to choose the native word above the unfamiliar. In some scientific circles, there is a sort of snobism for preferring a foreign word over the native word, for instance the preference for 'feeding *ad libitum*' over 'feeding to appetite'.

In using scientific terms one must always consider whether the circle of readers for which the work is intended will understand it. If not, explain or define it when you use it first. This is equally necessary when you coin or introduce a new term. Once a term is introduced, it is important to be consistent. Do not alternate between 'atomic energy' and 'nuclear energy', or between 'seed coat' and 'testa'. The argument that this is stylistically undesirable is certainly not true of technical terms in scientific or technical reports. The rule that a certain concept must always be expressed in the same term is much more important in these works. It will be clear that standardization of terms (nomenclature) is very important.

Every writer of reports must collaborate as much as possible in this standardization.

Some problems of standardization of terminology are covered by ISO Recommendations: R—704 Naming principles, R—860 International unification of concepts and terms, R—919 Guide for the preparation of classified vocabularies, R—1087 Vocabulary of terminology, R—1149 Layout of multilingual classified vocabularies, and Draft 1951 Lexicographical symbols particularly for use in classified defining vocabularies. Besides loan words from foreign languages, there are other words which do not fit well into modern reports. They may give an old-fashioned or pompous impression. This can sometimes happen, for instance, with use of 'biologic' instead of 'biological', 'whereby' or 'thereby' for 'by which' or 'by this', and 'whithersoever' for 'whatever'.

A report cannot, however, go to the other extreme of informal language. In speech we might say 'the cow needs 20 kilograms of hay a day'. But in writing we may have to say 'the cow's daily requirement of hay is 20 kilograms'. It is easy to see the difference between spoken and written language in lectures and papers presented at symposia. In a lecture one may not say, for instance, 'as mentioned above' and conversely when one publishes a paper that has been presented as a lecture, one does not begin with 'Ladies and gentlemen, I have pleasure today in addressing you on ...'.

There is a tendency not confined to science for words to increase in length. This is particularly true of words of Greek origin. There is also a tendency towards noun clusters. Often this kind of compound leads to difficulty with hyphenation. Horse fly, scanning electron microscope and dry matter content do not give any ambiguity; but then what does the writer mean by 'a large pig rearing house' or 'Goss's early embryogeny paper'? 'A man eating tiger' could be 'a man-eating tiger' or a man who has killed a tiger and is eating it. Such difficulties may be overcome by inserting hyphens or by consolidation, by inserting prepositions or by converting some of the nouns into adjectives. 'Polyacrylamide-gel-electrophoresis-demonstrated distribution' is 'distribution demonstrated by electrophoresis on polyacrylamide gel'. A 'common adjective ending' could mean either a 'common adjectival ending' or the 'ending of a common adjective'. But watch for ugly strings of a particular preposition such as the 'serrations of the mandible of the lesser-spotted godwit', which is better written the 'lesser-spotted godwit's mandibular serrations'.

Clearly, personal taste comes into this problem. One person tends to think a certain combination is long before another; the reader may be happier to struggle with a long compound word than with the long explanation.

7.6.3 Direct style

Usually the style of a piece of writing must be as direct as possible. This means that the active form of the verb is preferable to the passive, that the 'I' or 'we' form is preferable to the 'one' form or the passive. One could characterize this style as 'coming from the person'.

Although this is true for extensive articles, there is some debate about its desirability in scientific reports. It is common for the scientist to remain outside the report he is writing. He wishes the report to be objective and tries to avoid all subjective elements. One rarely finds in a report, for instance, 'I found the following data by analysis of ...', but usually 'The analysis of such and such gave the following data' or 'By analysis, the following data were found'. A not entirely satisfactory compromise is the 'we' form in which 'we' corresponds to 'one'. It is worthy of note that the style of scientific publications has changed in the course of time. Writings of Antoni van Leeuwenhoek (1632–1723), for instance, are mostly in the 'I' form and even Lorentz (1853–1928) shows this personal involvement.

Another point is the importance of avoiding hollow phrases. The Americans speak of 'fog' or 'blah blah'; Professor Stuiveling called them 'words of wool with a kernel of nothing'. An example comes from the Statement attached to the Dutch agricultural budget: 'Development of the grower's price level as second determinant factor for the gross value of agricultural production, shows in the arable and horticultural sector a clear effect of—under the effect of the drought situation—an increase in demand for Dutch agricultural and horticultural products.' Or in ordinary language: 'The demand for Dutch agricultural and horticultural products increased and their prices rose because of drought in other countries'.

7.6.4 Figurative speech

All languages are rich in words and phrases that are used in a transferred sense. For readers who are less well read, they can lead to misunderstanding or lack of understanding. In reports they should not be overused. One must take particular care if the figurative meaning is close to that of the words in normal use. A strange example is: 'Bottlenecks are growing up all over the country'.

7.6.5 Logical jumps

If one reads 'A dog flew suddenly through the open door and before I knew it I had a bite in the leg', one would not need to consider whether the bite came from a flea or the dog but the writer is still making a logical jump. Every report contains logical jumps. The question is only whether the logical jumps are acceptable or not. A simple method for a writer to check whether he is jumping too far is to insert mentally the words 'thus' or 'therefore' into the sequence. Often this will demonstrate whether the line of thought is still acceptable.

7.7 Readability research

There has been research on the question whether the reading ease of a publication could be measured, particularly in the United States. The background to this is clear: if one could make up a sort of formula whether a certain publication was difficult or easy to read, one would have a means of writing and rewriting a publication until it was easily readable by a certain circle of readers. Furthermore, one would know the factors that really improve the readability of a publication. A lot, however, depends on style, which has a personal character and is therefore difficult to measure or express in numbers. There is a comparison of different methods in Chall (s.a.). The book reviews many methods, all of which have advantages and disadvantages.

The best known method is that of Flesch (1962). This method is based on measuring the average number of words per sentence, the average number of syllables per 100 words and the human interest factor. These data are put into a formula based on experience. The result is a number that can be compared with the following scale (without human interest factor) which states the readability:

    0–30   very difficult
    30–50   difficult
    50–60   fairly difficult
    60–70   standard
    70–80   fairly easy
    80–90   easy
    90–100   very easy

Reading ease score (without human interest factor) = 206.835 − (1.015 × average sentence-length + 0.846 × number of syllables per 100 words).

It is interesting to apply this test for practice to a piece of one's own writing. But one can only expect an indication. It is advisable to aim for a text with sentences not averaging more than 24 words and with a maximum of not more than 40 words. To use the formula one sets to work as follows:

Take a random sample of ten lines, for instance from every three pages, of a work. Count the number of words on those ten lines and the number of complete sentences. A sentence ending with a semicolon counts as two sentences. Divide the number of words by the number of sentences to obtain the average number of words per sentence. Then make an average for all the parts sampled. For each part of the text take then 100 consecutive words and count the number of syllables (including words of one syllable) in 100 words. Then one can average the number of syllables per 100 words.

Another method is that of Taylor, called the Cloze procedure. This method is based on the theory that one tends to observe the whole and only afterwards the parts. If a part is lacking, one can fill it in more or less easily. A circle with a small segment missing, can easily be considered as a circle. If one sees, for instance, the sentence 'Hurry up; we are . . .' it is easy to fill in the word 'late'.

The closer the relationship between writer (text) and reader, the easier it is for the reader to fill in a number of intentionally omitted words (e.g. every tenth word). By this means one can test the readability of two different texts on a single group of readers. Conversely one can test one text on two groups of readers to gain an impression of the difference in readability for these groups.

7.8 Use of illustrations and tables

It commonly happens that one wishes to take illustrations (photographs, figures and graphs) from the literature for the literature survey. With modern equipment, they can easily be reproduced. Only the photographs give difficulties since, for instance, Xerox gives no satisfactory reproductions and one must make a photograph of them on film. Tables and graphs are reproducible on paper without difficulty. The making of graphs and tables by the writer of a report will not be dealt with here but under research report (Section 8.7.1 and 8.7.2). It is not usually necessary to replace letters and words when taking over tables and figures. The captions can be taken over too, but it is advisable to make one's own caption. One must therefore take care that all figures and tables in the report have their *own* explanation.

Photographs, drawings and graphs in the text are all called figures (Fig.). They are numbered in sequence through the whole report. *Each* figure must have a caption which should be sufficient explanation to make the figure understandable without the text. Only the interpretation of the figures should be included in the text. Make sure also that *all* figures are mentioned in the text. The same applies to tables. They too must be numbered in sequence (Table . . .) and must have a caption clearly explaining them. Take special care to mention units. It must not be left to the reader to guess the units in which a column of figures is expressed! An exception to the numbering and captioning of tables is a type of enumeration, small tables consisting of at most two columns which can be simply included in the text (after a colon). For instance:

'At the court of Gripsholm the recession was as follows:
in 1555 nutrition of 4166 cal (per person per day)
in 1638   ,,   ,, 2480 ,,   ,,
in 1653   ,,   ,, 2883 ,,   ,,
in 1661   ,,   ,, 2920 ,,   ,,
(From Slicher van Bath (1960), p. 94. Spectrum, Utrecht)'

7.9 References to the literature

There are three points that require attention in references to the literature:
(*i*) how do we cite the literature in the text;
(*ii*) how do we arrange the list of references or bibliography at the end of the report;
(*iii*) how do we describe the items of literature cited. (This last problem has been discussed in Section 6.2.4.)

7.9.1 Citing literature in the text

There are different methods of citing literature in the text. By far the most common are as follows:
(*i*) *Footnotes*. This method may be used only if there are few references and is thus not usually applicable for a literature survey. The notes are placed in the order of citation. Numbering should be in sequence through the whole report. Footnotes are troublesome for the typist.
(*ii*) *End-notes*. This method is used a lot by historians who give considerable explanation together with the references or sources.

This method is a nuisance for the reader who is obliged to refer to the end of the publication each time to see whether the notes hide something important. (Here, too, notes should be in the sequence they occur in the report and should be numbered in one sequence.)
(*iii*) *Numbers in brackets*. The numbers refer to the bibliographic descriptions in the list. The order of the numbers should be based on the alphabetic literature list and is thus not in the order of citation in the text. Literature lists that are not alphabetic do occur but that method is not advisable.

The advantage of the number system is that citation in the text is brief and does not disturb the account (particularly when a lot of literature must be cited in one place). The disadvantages are that even readers who know the literature well must still refer to the list; also, it is difficult to add new citations at a late stage because the numbers throughout the whole report must be amended. Fig. 7 is an example of citation in the text and of part of the list of references using this system.

In de jaren 1953—1964 slaagden enkele onderzoekers erin, haploïde weefsels te verkrijgen. In de periode 1964—1971 lukte het haploïde planten te verkrijgen van *Datura innoxia* (16), enkele *Nicotiana* spp. (7, 41, 54), rijst (17, 39) en *Brassica* (24).

Literatuur

16 Guha, S. & Maheshwari, S. C.: Cell division and differentiation of embryos in the pollen grains of *Datura* in vitro. *Nature 212* (1966) 97—98.
17 Guha, S., Iyer, R. D., Gupta, N. & Swaminathan, M. S.: Totipotency of gametic cells and the production of haploids in rice. *Current Sci. 39* (1970) 174—176.
18 Hackett, W. P. & Anderson, J. M.: Aseptic multiplication and maintenance of differentiated shoot tissue derived from shoot apices. *Proc. Am. Soc. Hort. Sci. 90* (1967) 365—369.
19 Heinz, D. J. & Nickell, L. G.: Separating and recovering genetic variability of a sugarcane clone asexually by use of cell and tissue culture techniques. XIth Int. Bot. Congr. Abstracts (1969) 88.

Fig. 7. Example of text citation via numbered references and example of corresponding list of references.

(*iv*) *Name-year system or Harvard system* with the author's name (without initials) and with the year of publication in the text between brackets, for instance (Greve, 1941). This avoids difficulties of Method (*iii*) but raises difficulty when many authors have to be mentioned together. It also raises difficulties in reference to work

101

without a personal author (corporate authors), which can, however, be manipulated with acronyms.

If there are two publications by an author in one year one may place letters a and b after the year or specify the month of publication by the ISO Numeric system (e.g. Schaal, 1940–01). Fig. 8 gives such an example using a and b.

Among isolates of *S. scabies* physiologic specialisation is described by de Bruyn (1939), Leach et al. (1939), Schaal (1940a, 1940b), Taylor & Decker (1946, 1947), Thomas (1947), Emilsson & Gustafsson (1953), Gregory & Vaisey (1956), Hoffmann (1954, 1959), Weber & Menzies (1962) and Mygind (1962).

References

Sanford, G. B., 1926. Some factors affecting the pathogenicity of Actinomyces scabies. Phytopathology 16: 525–547.
Schaal, L. A., 1934. Relation of the potato flea beetle to common scab infection of potatoes. J. agric. Res. 49: 251–258.
Schaal, L. A., 1940a. Cultural variation and physiological specialisation of Actinomyces scabies. Phytopathology 30: 21 (Abstr.).
Schaal, L. A., 1940b. Variation in the tolerance of certain physiologic races of Actinomyces scabies to hydrogen-ion concentration. Phytopathology 30: 699–700.

Fig. 8. Example citing author and year in the text and example of corresponding list of references.

For this system, it is necessary to place the year directly after the name or names of authors. This position deviates from that commonly used by librarians and documentalists (Section 6.2.4).

7.9.2 The literature list (bibliography)

The list of literature must be alphabetic. For the benefit of the reader subsidiary elements of the name not cited in the text, for instance initials, are placed after the *first* author's name for instance van der Mark, D. For further authors, the name is sometimes not inverted. Usually not more than three authors are mentioned. If there are more, they may be reduced to '*et al.*' (*et alii*). The type of letter used in bibliographic references is a matter of housestyle. The names of authors are commonly printed in small capitals (apart from the first letter of the name and initials), and the titles of books and journals are commonly printed in italic. (Small capitals are capitals the same height as normal lower case letters, for instance MARK, D.)

A difficulty is the link between successive authors. In different languages: and, en, und, et. They may be translated into the local language or printed in the original language, printed ampersand (&) or merely separated by semicolons (;) as in many abstract journals.

Unfortunately there is neither national nor international uniformity in how to quote bibliographic references. Lists in different publications vary widely. Even the alphabetic arrangement is not uniform. The British practice is, for example, to include J. van den Berg under V and not B, whereas in Dutch it is included under B. Chemists tend to give less information in literature lists than biologists; historians have a different system again. It has been calculated that there are 2632 ways of writing bibliographic references. Efforts at international rules are indeed in hand.

On one point, international agreement has been reached. New books are numbered by an international system, the International Standard Book Number. This number is placed on the back cover of new books and on the verso of the title page with the prefix ISBN. It is advisable to mention the ISBN when ordering a book from a bookshop or from the publisher.

Example
ISBN 90-220-0369-8. In this number, 90 stands for the country (Netherlands), 220 for the publisher (Pudoc), 0369 for the particular book (Labruyère, R. E. *Common Scab and its Control in Seedpotato Crops*) and 8 is a check digit.

7.10 Rounding off the report

Finally, a few short practical notes on the final version of the report:
   use standard sized sheets, preferably A4 format (21 × 29.7 cm);
   use good paper and type with a good ribbon;
   check carbon copies both for the quality of the text and figures and also for the quality of the paper;
   check carbon copies also for details and see that they correspond to the original particularly for corrections that have been made;
   begin each new chapter on a new page and as far as possible do not begin a section at the foot of the page (end a page with not less than three lines of text);
   number pages in sequence through the whole report including prelims and end-matter, and place numbers in the top outer corner of each page;
   use one sort of typewriter with at least eight-point letters (small

letters at least 1.4 mm high) and keep a constant number of lines per page;

use margins of at least 25 mm (40 mm on a binding edge), the right margin varying by not more than 5 per cent;

keep a broad head and foot clear of text on the pages, for instance 4 cm at the head and 3 cm at the foot;

number notes continuously and not independently for each page;

indent paragraphs by about five spaces; whatever the housestyle in print, the typed copy should be indented by about five typed spaces;

do not leave any extra space between paragraphs;

type the final report with $1\frac{1}{2}$ linespacing;

even in a typed report there are many ways of distinguishing grades of headings: centred in the page, flush to the left, displayed or in the line of text, spaced by use of an extra margin, underlined or not underlined, capital letter or small letter;

a general hint: do not pack the text too close together; do not use too many capitals and do not underline too much;

correction of a report is exacting work; to do it well, one must check in three ways:

(*i*) for the letter (language and typing errors) and also for form (uniformity of headings, blank space, captions, numbering etc.);

(*ii*) for style (for instance excessively long sentences, sentences that are not complete);

(*iii*) for content (nothing forgotten, logic, clarity).

The report must include the following:

(*i*) A title page (this is *not* the cover!). The title page must include: the title of the report, the name of the compiler, date of completion, if necessary address and telephone number of the compiler, and possibly the name of the client and any report coding and numbering.

(*ii*) A page with a synopsis immediately after the title page.

(*iii*) Table of contents listing the chapters, sections and appendices, and the pages on which they can be found. The table of contents is placed on a separate page after the page with the summary.

(*iv*) A page with 'Explanation' (Section 7.3).

(*v*) The actual content of the report divided into chapters, sections, or only into sections and subsections.

(*vi*) End-matter or appendices if any.

(*vii*) A letter of transmittal or—if this is not made—a preface which roughly corresponds to the letter of transmittal. The preface follows the table of contents on a separate page. It should be mentioned in the table of contents.

(*viii*) A cover on which should be at least the title of the report and the name of the compiler and preferably also the date of completion.

There are different ways of binding a report together:
  very simple: staples (objection: it does not open properly);
  a somewhat better way is to staple the report but not the cover which is pasted on;
  even better but difficult with a large format: to staple through the spine (for this, the sheets must be folded);
  rapid binders make the report even more difficult to open;
  ring binders are better but pages tend to tear out;
  plastic binders largely avoid this problem but storage of reports in this sort of packing is sometimes difficult. There are different types on the market;
  staple-less binding (Lumbeck binding), as long as this is well done it is the best solution;
  if there are inserts of another format other than the report itself, it may be necessary to provide the report with one or more pockets to hold them;
  remember that folded inserts on the *outside* (thus visible) must be provided with a description of the insert;
  it is better to cut the *bound* report clean, in other words to cut them off smoothly with a suitable guillotine.

7.11 Some standards and other guidelines

Many national and international institutions, organizations and committees are concerned with standardization. In other words they try to set standards or specifications for numerous matters, from estimating the bacterial count of milk to the pitch of music, to take a couple of examples.

Of course standardization has little purpose unless the standards are generally accepted and applied. It is therefore essential that users know what standards exist in their subject. A good help is therefore the catalogue of the National Standards Institution. Such catalogues often contain also recommendations of the International Organization for Standardization which are also available from national bodies. International standards have the letters ISO followed by a serial number (formerly ISO-R). Draft International Standards have the prefix **DIS** followed by the number. British Standards have the code BS, American ones ANSI (formerly ASA) and Indian ones IS.

Below follows a selection of standards which are important for report writing and publishing.

Standards and guidelines for different types of document

SERIALS

BS—2509 (1970)　　Presentation of serial publications, including periodicals (includes guidelines on preparation of papers for periodicals).

MONOGRAPHIC RESEARCH AND DEVELOPMENT REPORTS

World Federation of Engineering Organizations (1972).
　　　　　　　　　A guide on the preparation of scientific reports. UNESCO, Paris, 23 pp. Also available as ISO-TC46—1075.
BS—4811 (1972)　　Presentation of research and development reports.
US Committee on Scientific & Technical Information (1968).
　　　　　　　　　Guidelines to format standards for scientific and technical reports prepared by or for the Federal Government. Federal Council for Science & Technology, Washington, D.C. 20506. (PB-180600). Also available as ISO-TC46—1078.

PRINTER'S COPY

BS—1219 (1958)　　Recommendations for proof correction and copy preparation.

ENGINEERING AND OTHER DRAWINGS

BS—308 (1964)　　　Engineering drawing practice. With Suppl. 1.
BS—3939 (1966–)　　Graphical symbols for electrical power, telecommunications and electronics diagrams. 22 parts & 1 suppl. up to 1969.

PAGE SIZE

BS—1413 (1970)　　Page sizes for books.

TITLE PAGES

General
ISO-R—1086 (1969)　Title leaves of a book.
ANSI-Z39—15 (1971) Title leaves of a book.
BS—4719 (1971)　　Title leaves of a book.
IS—790 (1956)　　　General structure of preliminary pages of a book.

Specific parts
IS—791 (1956)   Half-title leaf of a book.
IS—792 (1964)   Title-page and back of title-page of a book.

Document numbering
ANSI-Z39 (draft)   Technical report numbering.
Unisist-ISDS (1973)   Guidelines for ISDS. UNESCO, Paris (ISDS-IC—2.2).
ANSI-Z39—9 (1971)   Identification number for serial publications. Also ISO-TC46-WG1—45.
ISO—2108 (1972)   International Standard Book Numbering (ISBN).
ANSI-Z39 (draft)   International Standard Book Numbering.
BS—4762 (1971)   Book numbering.

ABSTRACTS AND SUMMARIES

ANSI-Z39—14 (1971) Standards for writing abstracts.
ISO-DIS—214 (1974) Preparation of abstracts.

LIST OF CONTENTS

ISO-R—18 (1956)   Short contents list of periodicals and other documents. Under revision ISO-T46—1065.
IS—794 (1956)   Practice for table of contents.

SECTION NUMBERING

ISO-R—2195 (1972)   Numbering of divisions and subdivisions in written documents.

PARTICULAR TYPES OF WORK

Contributions to periodicals
ISO-R—215 (1961)   Presentation of contributions to periodicals. Under revision as ISO-TC46—1087.
Royal Society (1974) General notes on the preparation of scientific papers. Royal Society, London. Revised version.
ANSI-Z39—16 (1972)   Preparation of scientific papers for written or oral presentation.
UNESCO (1968)   Guide for the preparation of scientific papers for publications. UNESCO, Paris. 7 pp. (SC-MD—5).

Contributions to proceedings
BS—4446 (1969)   Presentation of conference proceedings.

NOMENCLATURE AND TERMINOLOGY

ISO-R—704 (1968)	Naming principles.
ISO-R—860 (1968)	International unification of concepts and terms.
ISO-R—1087 (1969)	Vocabulary of terminology.
BS—3669 (1963)	Recommendations for the selection, formation and definition of technical terms.

BIBLIOGRAPHIC DESCRIPTION

ISO-R—77 (1958)	Bibliographic references: essential elements. (Obsolete: R—690 is better).
ISO-R—690 (1968)	Bibliographic references: essential and supplementary elements.
ISO-TC46—1080 (1972)	International Standard Bibliographic Description for monographs. (Submitted by International Federation of Library Associations).
ISO-TC46—1072 (1972)	Bibliographic references: essential and complementary elements: patent specifications. (Also ISO-TC46—1097; 1972).
BS—1629 (1950)	Bibliographic references. (Obsolete). Amendment p. 1 1186, May 1951.
IS—796 (1966)	Glossary of cataloguing terms.
IS—2381 (1963)	Recommendations for bibliographical references.

Codes for languages and countries:

ISO-R—639 (1967)	Symbols for languages, countries and authorities.
ISO-DIS—3166 (1973)	Code for the representation of names of countries.
BS—3862 (1965)	Recommendations for symbols for languages, geographical areas and authorities.

Typical words in references:

ISO-R—832 (1968)	Abbreviation of typical words in bibliographical references.

GENERAL ABBREVIATIONS AND CODES

Rules for (abbreviated) periodical titles:

ISO—4 (1972)	International code for the abbreviation of titles of periodicals.
ANSI-Z39—5 (1970)	Abbreviation of titles of periodicals.
BS—4148 (1970)	Abbreviation of titles of periodicals. Part 1. Principles.

IS—18 (1949)	Abbreviation for titles of periodicals. Obsolete.
Unisist/ICSU-AB, (1970)	Working Group on Bibliographic Descriptions. International list of periodical title word abbreviations. Obtainable from: ICSU-AB, 17 rue Mirabeau, 75016 Paris 16e. To supersede ISO-R—833 (1968).

Transliteration:

Greek
ISO-R—843 (1968)	International system for the transliteration of Greek characters.
BS—2979 (1958)	Transliteration of Cyrillic and Greek characters.

Slavic Cyrillic
ISO-R—9 (1968)	International system for the transliteration of Slavic Cyrillic characters.
BS—2979 (1958)	(See above under Greek).

Arabic
ISO-R—2333 (1961)	International system for the transliteration of Arabic.
ISO-TC46—SC2 (1972)	Brief minutes ...
BS—4280 (1968)	Transliteration of Arabic characters.

Hebrew
ISO-R—259 (1962)	Transliteration of Hebrew.

Hebrew and Yiddish
ISO-TC46—SC2 (1972)	Brief Minutes.

Non-Slavic Cyrillic
DIS—2805 (1972?)	Transliterations of non-Slavic languages using Cyrillic characters.

Japanese
BSI—4012 (1972)	The romanization of Japanese.
ANSI-Z39—11 (1972)	System for romanization of Japanese. (Approved as DIS; see ISO-TC24—Sc2—25).

DATA PRESENTATION

BS—1957 (1953)   Presentation of numerical values (fineness of expression; rounding of numbers).

Statistical treatment:
ISO-R—645 (1967)   Statistical vocabulary and symbols: first series of terms and symbols. Part 1. Statistical vocabulary.

Dates:
ISO-R—2014 (1971)   Writing of calendar dates in all-numeric form.
ISO-R—2015 (1971)   Numbering of weeks.

Symbols for quantities and units:
ISO-R—31 (1958)   Quantities and units. 12 parts, some still in draft.
BS—1991 (1961–7)   Letter symbols, signs and abbreviations. 6 parts.

Système Internationale d'Unités (SI):
ISO-R—1000 (1969)   Rules for the use of units of the International System of Units and a selection of the decimal multiples and submultiples of the SI units.
BS—3763 (1970)   The International System of Units (SI).
BS Handbook 18 (1966) Metric standards for engineering.
BSI-PD—5686 (1972)   The use of SI units.
BSI-PD—6461 (1971)   Vocabulary of legal metrology: fundamental terms. (Transl. of Fr. document by International Organization of Legal Metrology).
BS—350 (1962–74)   Conversion factors and tables. 2 parts + 1 suppl.
International Bureau of Weights & Measures (1970)
     SI: the International System of Units. (Approved translation). HMSO, London.

Mathematical symbols:
See: ISO-R—31-Pt 11; BS—1991; Pt 1. 1967
     Letter symbols, signs and abbreviations.

ILLUSTRATIONS

ANSI-Y15—1 (1969)   Illustrations for publications and projection.

| ISO-R—710 | Graphical symbols for use in maps, plans and detailed geological cross-sections. 3 parts so far issued. |

INDEX

ANSI-Z39—4 (1969)	Basic criteria for indexes.
ISO-R—999 (1969)	Index of a publication.
BS—3700 (1964)	Recommendations for the preparation of indexes for books, periodicals and other publications.
IS—1275 (1958)	Rules for making alphabetical indexes.

Filing order:

BS—1749 (1969)	Alphabetical arrangement and the filing order of numerals and symbols.
ISO-TC46—1086 (1972)	Rules for alphabetical arrangement of words written in roman letters.
IS—382 (1952)	Practice for alphabetical arrangement.

TYPOGRAPHIC UNITS OF MEASURE

| BS—4786 (1972) | Metric typographic measurement. |

All standards should be ordered from the local National Standards Institution and not from the body publishing the standard.

7.12 Worked example of the systematic method (see 7.1.4)

Below follows the manner of composing this Section 7 (in the original (Dutch) version), taken as an example of composing a report in the systematic manner (Section 7.1.4).

First stage

(Notes with keywords)

Composition, writing (literature) survey. Literature survey as example.

Different methods. From just write down type to fully systematic.

Card system with abstracts forms good basis. Work through them and note in a sort of descriptor notation. Then arrange groups and classify.

Different criteria for arrangement. Chronological, historical, from theory to practice or vice versa. Systematic by section and then logical and summary of all facts and then logical, draw conclusions.

First the pros and then the cons or pro–con, pro–con.
Selection from material.
Principles of arrangement: theoretical, didactical, chronological, psychological, logical.
Basic rules:
Causes and consequences. Literature or one's own research.
Arrangement according to number of words.
Arrangement of chapter, section, subsection. Letters, decimal.
Style and language. Listing literature.
What is readability? Examples.
Factors governing readability: length of sentences, choice of words, active or passive, I, we, one. Compound words, figurative speech, grasshopper logic. Research on readability: Flesch, Cloze procedure, formula for English. Other parts of the report: tables and figures (only making them up), literature list, citation in the text, ISBN, UNISIST, ELSE.
Rounding off the report: how to type it, margins, arrangement, enumeration of parts of the report, method of binding, numbering, correction.
Standards!

Second stage

(Arrangement in paragraphs with provisional heads, arrangement of descriptors in each paragraph, deciding on the total number of words (11 000) and deciding on the number of words per paragraph)

	Number of words
Different methods of composition	1300
From intuition to completely systematic	
Then in the order: compulsive writing, the pieces method, methods of the inverted pyramid, systematic method	
Advantages and disadvantages	
Systematic treatment	700
Card system with abstracts is a good basis	
Work through them and put them into descriptor notation	
Arrange descriptor notations in groups	
Arrange the groups	
Put descriptors into the arrangement	
Decide the size of the report in number of words	
Divide the number of words between the sections	

Principles of arrangement 300
Form a firm scheme for each section
Theoretical principle
Logical principle
Chronological principle
Didactic principle
Psychological principle

Basic rules 1300
Cross-reference as little as possible, except at the beginning
 or from paragraph to paragraph
No conclusions if the evidence must still be given
Unless taken as a principle but then be consistent
Arrangement of pros and cons. One's own opinion
Selection
Rounding of the report
Explanation
Summary
Letter of transmittal

Technical arrangement 300
Chapters, sections, subsections
Roman, Arabic, letters or decimal
Not further than 1.1.1.1
Headings and text should be readable apart
The character of the headings should be uniform

Language usage 1200
Literature on English usage
Defining 'readability'
Examples

Factors in readability 1800
Length of sentences, not too long but also not too short
Choice of words, English, Anglo-Saxon or introduced words,
 officialese, compound words
Directness of style, active or passive, blah blah
Figurative speech, example: bottlenecks
Grasshopper logic, example of the dog biting

Readability research 800
Flesch and Chall
Cloze procedure

113

Use of figures and tables	500
Caption, legend and heading	
Table in the text	
References to the literature	1000
In the text, footnotes or end-notes, (figures), author–year, examples	
List of references, no standard, ISBN	
Rounding off the report	900
Advice on typing	
Correcting copy	
Enumeration of parts of the report	
Advice on binding	
Standards	900
ISO	
BS	
ANSI	———
IS (Indian Standard)	11 000
Example of systematic procedure	3.5 pp.

Further reading

Chall, Jeanne S.; s.a. *Readability: An Appraisal of Research and Application.* Ohio State University; Columbus.

Flesch, R. (1962). *The Art of Readable Writing.* Collier Books; New York.

Peterson, M. S. (1961). *Scientific Thinking and Scientific Writing.* Reinhold; New York.

United States Council of Biology Editors (1972). *CBE Style Manual*, 3rd edn. American Institute of Biological Sciences; Washington, D.C.

8 The research report

The research report has a distinct place among scientific publications. It is directly recognizable by its systematic composition. That does not mean that all research reports have the same form, but does mean that there are certain elements common to all research reports. This regular pattern is an advantage that should not be underestimated for the users (readers), because they can obtain a rapid view of what is being covered. Also if the user is only interested in part of the research (for example the methods used or the results) he can, by this regularity, rapidly find what is interesting for him.

The research report is of course based on the course of the research itself but does not need to follow it slavishly. In other words, the report does not need to follow the chronological order of the research project. The report should nevertheless follow the scheme forming the basis for the research and one could say that the report describes the briefest logical path one could follow from hypothesis to solution. One could express this as: ?———!

8.1 The scheme of research

To explain the scheme of a research report, one must, of course, begin with the scheme of research. As a starting point, we take applied natural science in which there is usually a clear scheme. How does such a research project come to be and what course does it take? Usually it begins here or there in the community where a difficulty arises which cannot be explained or solved. If such a difficulty arises only once, there is no reason to bring in research. But if the difficulty arises in several places or can be considered immediately serious, and when there is a clear general pattern, one can speak of a problem susceptible to research. Whether research begins or not depends on many factors such as the economic; social or political significance of the problem; the material, personal and financial scope of the research system; the costs attached to such research, and its urgency.

Industrial research in large firms carried out in their own research department originates along this same sort of line. Besides that type can be mentioned research on innovations intended for the working out of new ideas. This last type of research is also made use of generally, for instance in Great Britain by the Research Association and by the Agricultural Research Council, and in a more practical way by the agricultural advisory services or extension services. If the problem gives rise to research, this does not mean that the problem can be sharply formulated. Secondary effects or different conditions in certain places can, for instance, disturb the picture.

If one has recognized the (still vaguely described) phenomena or ideas as a problem and accepted it as an object of research, it can be put into the hands of the research institution or department, for instance through the management or administrative body that determines research programmes, with the request that solution to the problem be sought. The problem has then reached the stage of a *research project*.

Let us assume that a scientist has been asked to seek a solution. How will he set about it?

8.1.1 Collecting basic data

Initially he will try to collect as much basic data as possible about the still vaguely defined problem. Afterwards it will often be seen that he has collected irrelevant data or missed essential data. It is therefore often necessary to collect data again later, which can be a nuisance. One must therefore try to be as thorough as possible at first.

Even at the beginning, one can usually give the problem a name (which may not however be the right one); for instance: Planta disease, flat neurosis, nuclear fission, earth rays.

8.1.2 Literature searching

Simultaneously and later, the scientist must study the literature or have an information officer look up the literature in order to find out what is already known about the problem, and to avoid duplication of research. Usually he will find something or other that will reduce the problem to smaller proportions and reduce the scale of research. Let us assume that, despite this, the essence of the problem remains unresolved.

8.1.3 Scope

The remaining problem must now be defined as sharply as possible. This is the research scope or definition of the problem. It is a necessary preliminary to the research; we will return to it in Section 8.5. Until the research is complete, the scope remains part of the project description in which all data that are necessary for the research administration to assess the project are collected.

The project description serves also as a basis for progress reports. In general, it contains the following elements: title of the project, name of institution and department where the project is carried out, name of the project leader (and of those involved in it), scope of the project, manner in which it is planned to approach the project, phases in the research, a prediction of the time that will be required and an estimate of the costs. Project descriptions can be arranged by different criteria and can then serve as a basis for indexes of current research.

8.1.4 Working hypotheses

Then a number of working hypotheses are elaborated which could give a solution to the problem or to parts of it. For each working hypothesis, the scientist works out a method that can demonstrate its correctness (or otherwise) so that it can then become a conclusion or theory.

At this stage, the scientist sometimes begins a new, more thorough, literature search on one of the working hypotheses. From this, it may be possible partially to solve the problem from the literature and so to reduce the research project to a smaller scale.

8.1.5 The research

The research that now follows comprises three elements: setting up of trials, observation, and recording of data from the observations. When these activities are complete, the whole must be correlated and, after due consideration (and often many attempts), the results must be processed and put into a logical order so that they are capable of interpretation. These results can be recorded in the form of preliminary conclusions.

This general scheme can, of course, have all sorts of variants. It can be practical to have two working hypotheses in one research project; it can happen that a working hypothesis leads into a cul-de-sac; sometimes the method chosen gives no satisfactory answer to a hypothesis and the problem remains; a further literature search on parts of the problem may be desirable; a conclusion can be negative. The course of the research is given in Fig. 9.

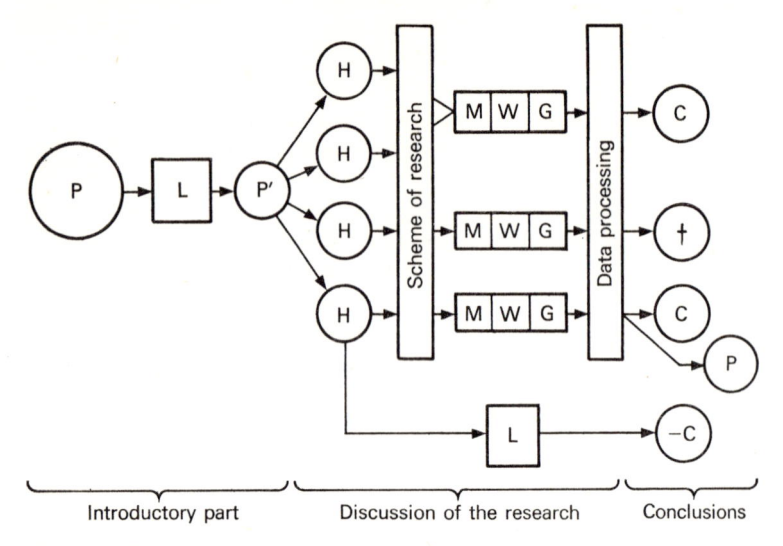

Fig. 9. Scheme of research and a research report. P—problem; L—literature search; P'—scope; H—working hypothesis; M—method of research; O—observations; D—recording data; C—conclusion; †—cul-de-sac; −C—negative conclusion.

8.2 Scheme of the research report

The scheme of the report can easily be derived from the scheme of research. We can distinguish three parts (Fig. 9):
 the introductory part;
 the discussion of one's own research;
 the conclusions.
The report includes also a fourth part: the appendix. In fact, there is even a fifth part: the title.
 We shall examine these parts more closely.

8.2.1 Introductory part

In this part, the problem must be defined. This can be included in a survey of how the problem was defined: *the cause for research*. In other words, one describes the background to the problem, how the problem came to light, where it has occurred, etc.

This survey is closed off with a clear formulation of the question or questions that must be answered in the report, in other words the *scope* or *definition of the problem*. Since scope forms the basis for the whole research (Section 8.1, Scope), the greatest care must be taken in formulating it in the report. It must be immediately clear to every trained reader what the problem is. The project description indeed gives a good first draft of this formulation.

If one has consulted the literature, this must be described in the introductory part before formulating the problem. If there is much literature, it is desirable to discuss it in a separate section, *review of the literature*. It is then advisable to have a separate section too for scope.

The way in which literature is traced, analysed and described corresponds to that used in the literature survey as discussed in previous chapters. The difference is more quantitative than qualitative. A literature survey is generally larger than a review of the literature in a research report. Sometimes one needs to discuss further literature in other places (for instance in the section on techniques). There is no great objection to this so long as it remains a simple short citation. It must be emphasized that no new literature may be cited or described in the conclusions section.

The introductory part thus consists of three sections: description of the reason for research, the review of the literature and the scope of the research.

8.2.2 Discussing the research

After this introductory part comes the discussion of the whole course of one's own research. This usually takes the bulk of the report. This part of the report must not be a chronicle or diary of research. It must likewise be built up logically and systematically. This does not, of course, mean that there is any harm in keeping a diary. On the contrary, the noting of operations carried out each day, the listing of observations and of data that one has obtained is an essential aid to memory in writing the report. The most obvious arrangement is a section describing the experiment as a whole, a section on methods in the experiment, a section on the results obtained (often in the form of numerical data) and finally a section discussing these results.

In *describing the research*, one must answer the question *what has been done* and *why*. This section will include the working hypotheses on which experiments were based, a description of materials used, where and how the trials were set up, the places where samples were taken and what chemical analyses were made. It is remarkable that many reports in the natural sciences contain little about the working hypotheses on which the research is based. Usually the introduction is immediately followed by a description of one's own research without any indication of why the research was carried out. In reports of research in the social sciences, however, this section on the working hypotheses is often extensive.

Methods of research do not often need to be described fully. If they are known methods described earlier, it is sufficient to mention the literature where they are described. As always, one must here cite the original publication; one cannot use a derived description in a textbook as a basis. Sometimes there are many variants of a particular method. One must of course clearly state which variant one is using. If one has invented a method or made a variant for one's own research, it must be carefully described. The description must be adequate for another scientist to do it *merely by following the description*. There must be no doubt about the time taken, temperatures, units, weights, chemical materials etc. For this reason one must adhere to (national or international) standards of nomenclature.

The *results obtained* in the form of recorded observations will only be partly in words. They will usually be given in the form of tables or figures (Section 8.7).

Discussion of results is perhaps the most critical part of the whole report. So far, it has been sufficient merely to describe. In this part one must use reason and logic; here the scientist is put to the test. Obviously one scientist will draw more information from his research than another. It depends on intuition, inventiveness, background knowledge and perseverance, in short on the human qualities of the scientist. In Section 8.6 we will go deeper into this.

This part can most easily be described if one has first attempted to formulate (tentative) conclusions. This forms a basis and one can divide discussion of results into subsections or paragraphs by analogy with the conclusions. In this part, one needs to go back into the previous section on results. It is sometimes indeed practical to combine these sections. Otherwise one may merely put the results of observations (for instance in tables) in the section on results and the (mathematical) processing of these results and demonstrational figures in the discussion of results.

If an experiment consists of more than one part, one must decide on the order of the different sections, on 'results obtained', and 'discussion of results'. These sections can be arranged in a series by part of the trial or by working hypotheses or factor, or firstly all observations can be summed up and then the discussion of them all can be combined. The advantage of the former method is that the description is easier to survey: the parts remain close together. The advantage of the second method is that combinations of factors can be more easily dealt with. The report forms a more coherent whole.

8.2.3 Conclusions

Formulation of definite conclusions is simple if one has been considering them while writing the discussion of results. Conclusions then follow logically from previous paragraphs and it is merely a question of arranging them point by point. This last section of the actual report can therefore be short. It should not be used to re-open discussion and certainly not to discuss further literature! The term 'conclusions' must not be too restrictive. As already stated, a negative result can be a conclusion. But there can also be points in the conclusions on which doubt remains ('The experiment does not clarify whether .... Further research on this would be useful.').

It is sometimes useful to add something after the short section of conclusions. This can be so for instance in research that has practical implications, as happens, of course, more often in applied research than in fundamental research. In such a case, a need will be felt for a sort of afterthought to report various matters. Sometimes such a section may even be entitled an epilogue. Occasionally such a section may be necessary for a fuller treatment of the theoretical basis. A further separate section again is the 'word of thanks' to colleagues or acknowledgments which form a small section after the conclusions. But acknowledgments are sometimes alternatively put into the first part of the introduction.

8.2.4 End-matter

The report closes with end-matter consisting of two or perhaps three parts, the summary (in one or more languages), the literature list (or bibliography) and perhaps appendices.

The position of the *summary* is less certain than it used to be. As in unpublished reports, it is becoming more usual in journals for the summary directly to follow the heading of the article. For the documentalist it is indeed more logical if the title, author and summary are close together. Sometimes the bibliographic description of the articles and their summaries are collected (as abstracts) on a separate sheet at the front of each issue of a journal. This documentation sheet can be printed on transparent paper making it easier to reproduce on light-sensitive cards. Sometimes the sheet consists of cards that can be torn out or cut out.

It must be stressed that the summary is not the same as conclusions. Conclusions express in words no more and no less than what the research has achieved. The summary should reproduce the whole content of the research report. There are still many scientists who combine these two parts of the report into a final section entitled 'summary and conclusions'. They misunderstand the function of summary and the sharp formulation of conclusions is endangered.

Section 6.3 has described how to prepare an abstract and the same procedure applies to summaries. Whether one may write an indicative or informative abstract sometimes depends on journal policy. If your choice is open, an informative abstract is preferable to an annotation or to an indicative abstract. One must remember that the summary of a paper may have an independent life as an abstract. It acts as a birth notice for the whole paper. In the abstracting journal, this birth notice comes under the eyes of colleagues all over the world. One can thereby be assured that even this birth notice will be examined very critically.

At the end of the abstract, or instead of the text of the abstract, may be placed a list of descriptors or key-words. Descriptors are terms that may be used in indexing the report. They may be freely selected from the content of the summary or the text of the article, or may be taken from the thesaurus of a major abstracting service. There is no need for the list to include words already listed in the title.

The *bibliography* can be compiled in the same way as for a literature survey (Section 7.9.2). If the research report is to be published in a journal, it is advisable to read the instructions or guide for authors in the journal. Unfortunately there is no uniformity in the style of references. Many journals include instructions for authors in each number, often on the inner cover. If not, they can be requested from the journal editor.

Whether information be included in *appendices* or in the text is arbitrary. Some examples of parts of a report that may be considered for inclusion in appendices are:
    long tables that would interrupt the text;
    tables that do not fit into the normal page (fold-outs);
    maps of larger size than the page and coloured maps;
    figures or photographs that must be printed in colour;
    photographs that must be printed on art paper unless the whole journal is on art paper;
    transparencies that must be laid over figures.

8.2.5 Title

There is good reason for leaving the title until last. Although even when the research is starting a tentative title must be used, it is useful to wait until after the report is written to formulate the title. Title, definition of the problem and conclusions must correspond.

For the title there are two requirements:

it must be short for easy scanning;

it must give the content of the report as accurately as possible.

These two demands pose difficulties for the author. A precise title is sometimes quite long. But for machine systems of information retrieval authors tend to pack as many words as possible into the title to ensure that their material is retrieved. Even for machines there seems to be a limit of about 150 characters (letters + spaces). For the reader, however, a title of even 100 characters is too long. An incomplete, inaccurate or catchy title such as 'A matter of Life and Death' (the title of the proceedings of a symposium on the dangers of pesticides), can cause a publication to be lost in documentation systems or even to be put into the wrong category. If, for instance, research has been done on the effect of dressing grassland with copper salts on copper deficiency in milking cows, one may not use the title 'Copper dressing of grassland' nor 'Copper deficiency in cattle'. One must take for instance 'Effect of dressing with copper salts on copper deficiency in milking cows' or 'Pasture dressing with copper salts and copper deficiency in milking cows', or perhaps in a farming journal 'Dressing with copper salts alleviates copper deficiency in milking cows'. Although aiming for completeness, one must avoid embroidery such as 'Comments on . . .', Some experiments on . . .' or 'Contributions to . . .'.

Below the title come the names of the author or authors and their working addresses. The author's institutional address is important for questions or for ordering offprints. It should be sufficient for postal purposes. If there is more than one author, there are difficulties in deciding on the order: according to contribution to the research, alphabetical, or according to their seniority. There are no rules. Traditionally the author was merely the writer of a report irrespective of who did the research.

It is not usual to give authors' titles, for instance no Dr. or Mr. At least one forename is often given in full to avoid confusions with like-named authors. One may, however, give a short description of an author's function or a short biography in a footnote to the title page.

Often the title and author line are immediately followed by a short description of the content, sometimes called an extended title.

Sometimes the subject class code for example in the Universal Decimal Classification may be given here; then an abstract and a list of descriptors.

8.3 General rules

The general scheme of a research report thus comes to look as follows:
   Title and author line
   Abstract in an international language
   Introduction
      Reason for the research
      Review of the literature
      Scope (of the research)
   One's own research
      Description of the experiment
      Methods used
      Results obtained
      Discussion of the results
   Conclusions
   Abridged translation (for instance an English translation of a Dutch report) or abstract if these are not placed after the title
   Bibliography
   Any appendices.

One must never start writing a research report without establishing a scheme for it and deciding how large the report is to be (in number of pages or words). The number of pages or words must then be divided between the sections. The way of doing this is described in Section 7.1.

The basis for the report must be defining the problem and answering the problem (the conclusions). This is the only way of obtaining a good logically arranged report. The research report is very often the only result of research. An experiment is not complete until the results have been written up. It is the last part of research and must be budgeted for as part of the research.

The research report must justify the research but it must also be a condensation. The report must, furthermore, provide complete information about the research. Close attention should be paid to the making of the report and the following rules must be remembered:

The report must tell the truth. It may not be suggestive or tendentious.

One may not, for instance, omit facts that are at variance with the conclusions. The researcher/author has a doubtful predicament when, by error or accident, a result or a number of results are lacking or are very inaccurate: whether simply to deny these results or to describe a fault.

The report must be objective. Subjective elements such as feelings, and personal, intuitive or impulsive opinions must be avoided. But it is unavoidable that every piece of research—and also every report on it—is personal. The subjective element is present, particularly in the working hypotheses.

The report must be documentary, in other words it must contain all the information necessary for someone to repeat the research.

The report must contain only the essential. This rule is at variance with the previous one. The scientist as author must choose between completeness and essentiality.

The report must be constructed logically, even if the experiment—viewed in retrospect—does not follow logical lines. The report does not usually give a day-to-day survey of events. The diary must therefore be transposed.

The report must meet standards in the scheme, the arrangement, the tables, figures, nomenclature and bibliography.

8.4 Report of research on a method of analysis

A rather divergent type of research is into methods of analysis, in other words finding the best method of estimating a certain quantity or quality. For completeness, a scheme of such a report is given below.

(*i*) Reason for seeking a new method of analysis.

(*ii*) Defining the problem: determining the requirements that the method must satisfy:

(*a*) The method must supply relevant data.

(*b*) The method must work constantly over a certain range in certain conditions, various quantities, and differences in the materials to be analysed.

(*c*) The method must be easy to do.

(*d*) The method must be suitable for automation.

(*iii*) Description of existing methods (citing the literature) and experience with these methods.

Advantages and disadvantages of these methods.

(*iv*) Establishing the standard with which the method to be developed will be compared.

(v) One's own research: modifying existing methods or developing an entirely new method.
Describing difficulties during the trials and how the difficulties were avoided or overcome.

(vi) Check on the suitability of the method developed for use on different objects and comparison with the standard. Comparison between duplicate estimates.

(vii) Evidence for the correctness of the new method, statement of the limits within which the method can be used, conditions and peculiarities that must be considered.

(viii) Accurate description of the method elaborated.

(ix) Summary (if not placed immediately after the heading).

(x) Bibliography.

(xi) Appendices: results obtained in tests with the method and deviation from the standard.

8.5 The scope: defining the problem

The report must, to the exclusion of all else, be directed to answering the problem and must provide a coherent answer to the problem. Conclusions must not be drawn to questions that are not framed in the scope note and likewise all questions that are framed in the scope note must be presented in the conclusions (Fig. 10). Formulation of the scope note immediately raises the difficulty that the problem may not even be clear when the research is started. It may even be that a certain phenomenon is not viewed as a problem.

For example: the simple fact that potatoes flower and bear berries only once a year in temperate climates so that breeding by seed can only be slow, does not turn the harvesting of seed from

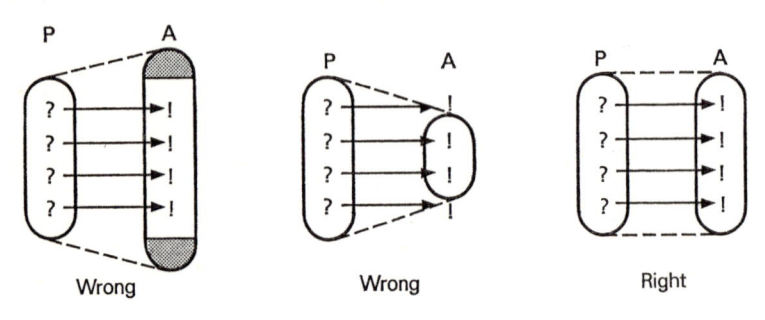

Fig. 10. Scheme of the definition of the problem and conclusions. P—problem; A—answer.

potatoes into a research problem. Only by such a formulation as 'Is there a method by which potatoes can be made to bear berries more than once a year in temperate climates?' can the problem be tackled by research.

An important job for the creative scientist is just this tracing of problems suitable for research. Scientists with a creative turn of mind can see such problems more easily than scientists with a systematic bent. One cannot learn this sort of vision but, if possessed, it can be cultivated by exercising scientific thinking and by building up a broad scientific knowledge. The education of the science student must be directed largely towards this. This inclination develops not only by getting to know large numbers of facts (even though this is certainly important) but rather by learning to think out the principles of one's own discipline and of other disciplines (Crombie, 1961). As science develops, there is a growing interdependence between disciplines and it is because of this interdependence that fruitful work is emerging. Indeed, in this way new branches of science have developed such as biochemistry, econometry, anthropography and agrometeorology.

For example: The search for an answer to the question as to whether acoustics could be applied in soil science has a typical interdisciplinary character (Janse, 1969). In examining research reports, one can see different types of scope note in which there are indeed transitions from one type to another.

Let us examine some types that occur in scientific research.

8.5.1 The scope note in applied science and development

In applied science, the problem often crops up on its own. If, for instance, the egg trade complains about the number of breakages, it is obvious to ask: 'What are the causes for frequent breakage of eggs?'. If this question can be satisfactorily answered, the question would then follow: 'What means or measures can be used to remove these causes?'. But this simplicity is only apparent. It would of course be quite wrong to start with an extensive study of factors influencing shell thickness of eggs before finding the answers to various questions such as:

Is there statistical evidence that breakage is more frequent in eggs now than earlier (for instance before World War II)?

Is there evidence that changes in the way eggs are handled in the trade could cause the problem?

If the answer to the first question were confirmatory and the second proved wrong, further questions could be asked such as:

Has there been research on this subject earlier and, if so, is there any trend to be found in it?
Are there changes in poultry husbandry that could cause the trouble?
And this question could be split up into the following:
other commercial breeds or crosses?
higher production per hen?
other methods of housing or management?
other feed mixtures?
a difference in shape of the eggs (for instance more tapered)?

Each of these questions could form a separate problem and research could be based thereon. The question whether there was earlier research could, of course, lead to a literature survey and would certainly have to precede study of the ensuing questions. It is then the task of the scientist and also of the management of his institution so to choose the problem that the research does not sink into a bottomless pit and, on the other hand, does not become so detailed that one cannot see the wood for the trees.

In other ways, too, the framing of questions in applied research is not simple, in particular if the phenomena are various. One receives conflicting reports from the farmer or from industry so that it becomes difficult to sort out the real problem. One is frequently dealing with secondary causes, which disturb the primary causes of the phenomenon. Here, too, the scientist has a difficult job eliminating disturbing elements or suggestions which are irrelevant, in order to reach a workable problem. For his research he must often then choose the reverse: he tries to induce the phenomenon under known conditions.

The back of the work is often broken when the question can be well framed. The solution can then often be found by traditional working methods of that type of research. One often knows the solution in advance and must find the way of reaching this solution. By analogy with the object of research, one can take a similar one that has already been solved. One can also start from the same solution as required but found for another object. By projecting research methods that are used in comparable cases, the question at hand can be solved. This is expressed in the scheme of Fig. 11.

8.5.2 The scope note in pure science

There have been exhaustive discussions on whether one can speak of a distinction between fundamental and applied research. What has become clear from these discussions is that if there be a distinction, it is anyhow less than was assumed in the past. There is a transition rather than a sharp difference.

The definition of the problem is usually slightly different in fundamental research. The problem does not arise of itself but is posed by the scientist himself. One could say: he does it out of curiosity without immediately thinking of any practical application.

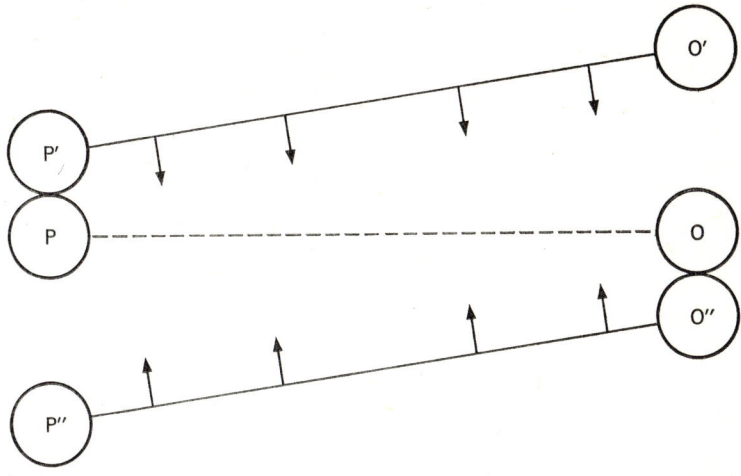

Fig. 11. Scheme of applied scientific research. P—the problem to be solved; O—the solution; P'—analogous problem with the same object as P; O'—the solution found; P''—problem with another object; O''—analogous solution to O; continuous line—way found; broken line—the unknown way; arrows—projection of the method.

Example: If we do not see anything special in the falling of an object and consider it as obvious, we will not start searching for the laws of gravity (just as we consider it obvious that one person is a better poet than another). Who can say how many interesting problems are simply not seen in the myriads of phenomena in our environment.

A remarkable fact is that one can even speak of a type of vogue phenomenon. One finds certain subjects that have attracted attention of scientists for a shorter or longer period, to be forgotten and then, perhaps later, to re-appear on the agenda again.

Example: The question of a fundamental material from which all other material is constructed has intrigued men since the dawn of time. Between 600 and 300 B.C., the Greeks developed many theories about what is now called nuclear physics. The term 'atom' derives from Laisippos (died 475 B.C.) and was taken up by Demokritos ($\pm$ 460–380/370 B.C.).

Alchemy was in fact based on the same theory. Its high point came in the 13th and 14th century; interest then died away. Only at the beginning of this century was research started again and fundamental research on material is again in vogue. One has the impression, however, that the vogue is on the decline. Thus, there is no question of systematic surveillance of the unknown areas of science.

Another question now arises of whether every problem that someone recognizes is worth being studied. Or, in other words: what is scientifically an interesting problem? This question cannot unfortunately be answered definitely. Every phenomenon in nature or in the community, however simple on first sight, seems on closer examination (that is in asking how and why, or in seeking the causality) complicated, and anyone who wishes can turn it into a long scientific study. One could say that all that is worth knowing is suitable for scientific research, but this merely shifts the difficulty. What one person thinks worth knowing is hardly worth the study according to another. It often depends on how the problem is tackled.

Example: the body measurements of a woman are worth knowing for the clothes industry. But one would hesitate to set them up for scientific research! Yet there has been research on human body measurements that is undoubtedly scientific (Sittig & Freudenthal, 1951). There is even a branch of science that deals with it: anthropometry.

On the other hand, one can ask whether a study on digestibility of cacao pods by milking cows belong to scientific research in the strict sense. The information could be important for dairy farming and the method of research would undoubtedly be scientific. Yet there is a clear difference between the scientific value of this information and the knowledge of the principles of cell division. The complexity or simplicity of a problem is in itself no criterion of the scientific character of research. Viewed for its universality, there is no question of complexity or simplicity. The construction of the atom is as 'natural' as the fall of a leaf from a tree. Only for the way in which men can observe it or conceive it can the one phenomenon be more difficult to understand than the other. The only criterion can be that science attempts to formulate laws in *as general terms as possible.*

Example: Seeking the laws behind absorption of *certain* nutritive elements from a *certain* soil under *certain* conditions of climate and weather by *certain* plants is scientifically less interesting than seeking the laws behind the absorption of nutritive elements from the soil by plants *in general.*

There is, however, a conflict: on the one hand one tries to frame the problem as narrowly as possible to solve it within a reasonable time; on the other hand one attempts to broaden the problem as much as possible so that conclusions can be formulated as broadly as possible. A start is often therefore made with a limited problem and its applicability is then studied in broader terms. In nature phenomena are very complex. Fundamental research often first attempts to remove as many complications as possible and to reduce the problem to its simplest form.

Example: Working with a phytotron in which all the factors for growth of plants are kept constant except one. Afterwards successively more factors are introduced. In the beginning one thus works with a *model* that in fact does not correspond with reality. Often the natural object is even entirely replaced by the artificial model.

Example: The flow of water in the ground may be simulated by electric current; the working of a ploughshare in the soil may be simulated by a model on a scale 1:10 in a 'soil box' filled with a homogeneous artificial mixture of soil. It is well to remember that it is constantly necessary to test the applicability of what has been found with the model against reality. Fig. 12 illustrates this procedure schematically.

In fundamental research it sometimes happens that the scientist finishes up in a different place from where he intended, but never-

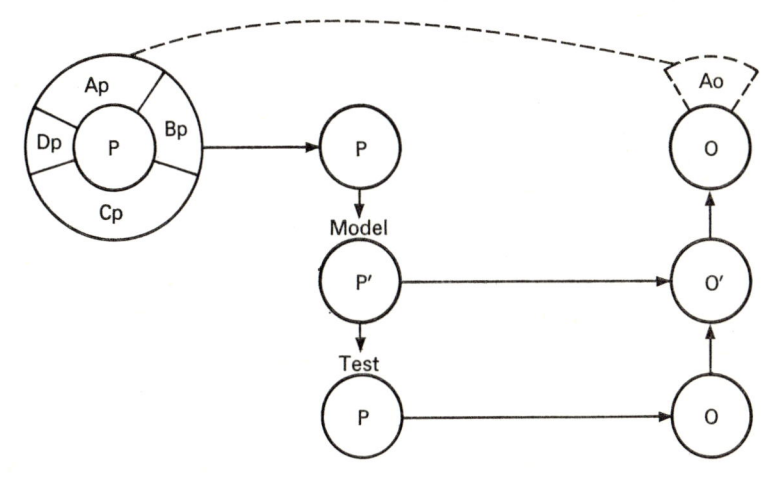

Fig. 12. Scheme of fundamental scientific research. P—unravelled key problem; O—solution to the key problem; Ap, Bp, Cp, Dp—other factors in the complex problem; Ao—solution to the factor Ap; P'—problem in the model; O'—solution in the model.

theless gathers important results. Since his report must not be devious, he is forced after his discovery to change the question he had framed. This is a sort of feedback as it is known in cybernetics.

8.5.3. Definition of the problem in trial and error

Between the two sorts of research comes trial and error. The problem is least clearly formulated in this. It is usually very vague or even lacking.

Example: It is discovered that the mould *Gibberella fujikuroi* causes plant cells to stretch. This action proves to be caused by an acid, gibberellic acid. The question now is: what practical applications can we find for this acid in agriculture? A common procedure is 'just to try something'. There are always results and, from the multiplicity of choices, there will certainly be a few of practical importance. An associated question is to make such an application possible on an economic scale.

Of course 'just to try something' is not the final word. One has some indication, perhaps vaguely, which way to seek for the greatest success. In applied chemistry, this method is often used. One must, however, remember that what is found in the laboratory is certainly not suitable for industrial application. The boiling of a liquid in a flask, for instance, is entirely different from the boiling of thousands of litres of the same fluid in a tank. This is called 'research *and*

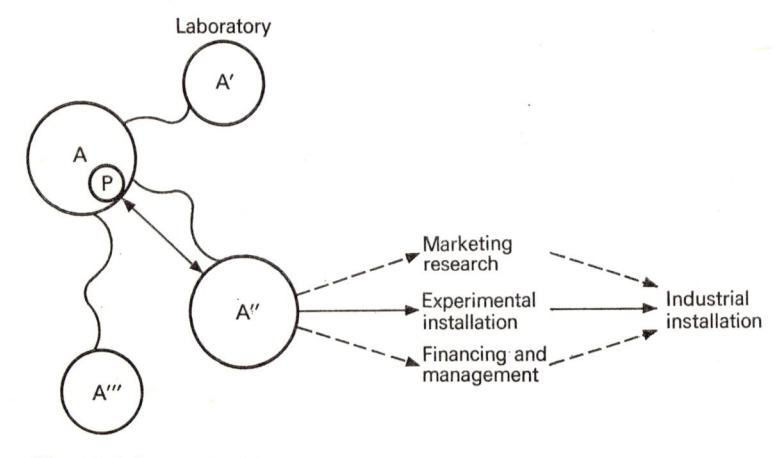

Fig. 13 Scheme of trial and error research. A—basis; A', A", A'''—derivatives found in which A" proves to have interesting properties; P↔—problem defined in retrospect in which the interesting properties are used as a problem.

development' (abbreviated R & D). After laboratory experiments there are often trials on a semi-industrial scale, then going over to an experimental factory. Only after marketing research proves satisfactory and all the technical and financial difficulties have been solved does production start on an industrial basis. The scheme of 'trial and error' research is illustrated in Fig. 13.

8.5.4 Summary

Summarizing, we can distinguish the following in defining the problem for scientific research.
(*i*) In applied research, starting point and purpose are certain. The defining of the problem gives no difficulty at first. The unknown is the path that must be trodden to reach the end.
(*ii*) In fundamental research, the scientist himself creates the problem. He must so choose the problem that he has a reasonable chance of finding a solution. His aim must be at finding laws as widely applicable as possible. Often he will have to change the question he is asking during or after the experiment (feedback).
(*iii*) In trial-and-error research, there is no sharply defined problem initially. This can be framed afterwards and certainly will need to be framed if an application is to be made practical (R & D).

8.5.5 Definition of the problem in sociological research

In a public lecture van der Zouwen (1971) reviewed the different types of definition of the problem that can be recognized in sociological research. He distinguishes the following types of problem:
The exploratory type: the question whether a phenomenon occurred and how often it occurred.
The descriptive type: research on a phenomenon over a period and in different areas.
The explanatory type: the question on causes and consequences of a phenomenon and on functions and dysfunctions.
The testing type: the testing of the correctness of opinions about a phenomenon. In assessing whether a question or sociological problem is indeed susceptible to research, van der Zouwen proposes two main criteria of relevance: the social and the theoretical relevance of the problem. One could compare this distinction with that between applied scientific research and fundamental research.

A question is *socially relevant* when it can be expected that the results of the research framed resulting from this question will contribute to the detection and solution of social problems. A question is *theoretically relevant* when its answer would contribute to the 'hypothetic-deductive' manner of forming theories. This type of theorizing aims at a coherent system of expression on the relations between phenomena in social reality.

8.5.6 The pseudo-problem

Pseudo-problems must of course be guarded against. A pseudo-problem can arise through incorrect or inaccurate observation (example: spontaneous generation) or through an incorrect premise (example: presence of an undetectable material, 'ether', in space, the existence of ghosts). It can be years before it is discovered that there is only a pseudo-problem. The task of disproving the existence of something is particularly difficult (example: earth rays).

8.6 The scientist's train of thought in the creative stage of research

After defining the problem the scientist must decide for himself how he will tackle the problem in a scientific manner. To the outsider this phase of research is hardly spectacular: nothing happens or at least very little. But, for the scientist himself, it is perhaps the most important phase. The choosing of a starting point that afterwards proves to be wrong, the following of a wrong method of analysis, the missing of a connexion with research by others can have serious consequences. He comes to a standstill with his research and must, years later sometimes, conclude that he has not achieved anything. The 'thinking work' preceding the actual research is thus very important.

This section will try to indicate some lines along which the scientist thinks in setting up his programme of research. As an example, we will take the natural sciences. Parallels with research of other types can be easily seen. One can distinguish research into that
    which exists or has existed (the existing),
    which will be (the predictable).

8.6.1 Research on what exists

Assuming that what exists in man's experience corresponds with reality, the scientist tries to penetrate the reality of the thing and to extract information. The extracting of information is dependent on what can be measured. The scientist tries, therefore, to push the frontiers of the measurable towards the invisibly small and the cosmically large. He thus comes beyond that which can be imagined, but which can still be formulated (the nanometre (= 10^{-9} m); the light-year (= 9460.5 Tm = 9.4605×10^{15} m). However, the frontiers are approached even of that which can be conceived: the fourth dimension, the expanding universe. At some points, we already reach something that is scarcely conceivable. By extrapolation, we can conclude that there are objects inconceivable to humanity, realities that mankind in principle and *de facto* cannot fathom. Objects that are inconceivable can clearly not be considered for research.

It may seem that men can reach further with modern computers. But the basis of the computer too is human thinking and the computer can do its work only within human limits. At the beginning of the 20th century scientists believed a complete knowledge of the world was possible. This belief is now shaky. There is doubt about one of the axioms of logic, *tertium non datur*: something is right or wrong, there is no third possibility. The certain or normative sciences in which something is true or untrue have been replaced by the science of the probable in which something is probably true, but there is a chance that it is untrue. Scientists no longer look for incontrovertible evidence but only for the relation between the objects, the correlation.

There is even a tendency towards the science of consideration: the relation between man and object (for instance: '*I consider* this or that to be true, probably true or untrue'). Science is aware that man can only study his relation with the object of study so that there is an interrelation between object and subject. In this way, science has returned to subjectivity as it existed earlier (for instance in the Middle Ages), though on another basis: with man considering himself as centre of the universe.

8.6.2 Research on the predictable

Applied science is largely directed to discovering the predictable. It is of no interest to discover that at a certain moment $A + B \rightarrow C$, but it is interesting to know that $A + B \rightarrow C$ always and everywhere.

One can distinguish between two starting points for research on the predictable

(*i*) Taking what is known as a starting point and trying, often by trial and error, (Section 8.5.3) to reach something new.

Example: The discovery of the grafting of potatoes onto tomatoes as a method of breeding potatoes. This method originated from attempts to show the influence of phenotype on genotype.

(*ii*) One takes the end as the starting point and tries, often by analogy, to reach this end.
Example: The breeding of a potato cultivar not susceptible to scab.

There are characteristic differences between research of Type (*i*) and Type (*ii*). In the first the emphasis is on the ability of the equipment to distinguish between the recording of data and its interpretation. In the second, the emphasis is on analysis of the end in view, the splitting up of the end into components and the reaching of it step by step. For both types of research there is a need, at a certain moment during the research, for retrospective consideration of the methods used in order to replace the game of science by a well-founded technique. Here applied science closely approaches the method of fundamental science.

In fundamental (pure) science one attempts to separate the object of study from the complexity of everything (Section 8.5.2). One eliminates all disturbing factors until the problem in its simplest form is susceptible to study. When results have been obtained, the disturbing factors are introduced one by one (Fig. 12). Often, however, some artificiality remains or the scientist studies a model in which he is aware that he is only partly approaching reality. Applied sciences often purposefully make use of complexity. Through this, applied sciences need cross-connections between different disciplines. Examples are disciplines like biochemistry, econometrics, physical chemistry, social psychology and soil mechanics. The whole system of the sciences is no longer two-dimensional but multi-dimensional. The recognition of this multi-dimensionality of the sciences leads to particular difficulties in the classification of literature, thus in documentation. The classification systems so far used are, at most, two-dimensional; they work on the horizontal level. Documentalists are therefore looking for other solutions. It is remarkable that in the mechanization of documentation, use is actually made of *one* dimension in which often powers of 2 (the binary system) are used. The machine continually chooses between two.

The old representation of the line the scientist takes in his research is that he has an almost straight path before him and progresses from milestone to milestone (Fig. 14). This representation proves incorrect. Moles (1957) gives an entirely different approach. He suggests that the scientist is in an unknown area. In that area there is a sort of labyrinth with numerous paths. The scientist knows only the 'vehicle', in other words the equipment and the working methods that are available and the laws of logic. But whither a certain way leads he does not know. Each way does, however, lead to something. But the question is whether this 'something' is worth knowing, even more so if one considers that the scientist is only really studying *his relation with* the object. As discussed in Section 8.5.2, the question whether something is worth knowing (in other words, is or is not scientific) is particularly difficult to answer. The answer can sometimes only be given years later.

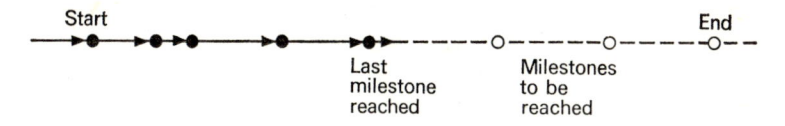

Fig. 14. Classical representation of the line the scientist takes (after Moles).

In the labyrinth, the paths continually fork. At each fork, the scientist has to choose which way to continue. In discovering ways in the unknown area, we have not one possibility but many. In seeking his way through the labyrinth, the scientist projects his thoughts on completed science, in other words the science already recorded in writing or publications. In this, he will take particular notice of laws discovered and will complete them with facts found later. As guides from these laws and facts, he will choose those that suit him, in other words those which have become part of him (Fig. 15). His way through the labyrinth is clearly determined by repeatedly choosing one of two or more possibilities and thus rejecting

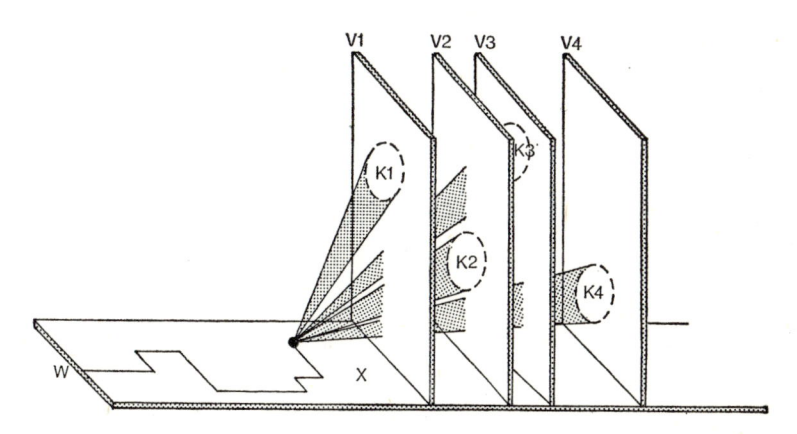

Fig. 15. The scientist's process of thought according to Moles 1957. W—the scientist's route; X—the unknown area; K1–4—the projection of his knowledge; V1–4—the various disciplines.

all other possibilities. Only exceptionally will the scientist follow two paths simultaneously knowing that one of the two paths will not lead to the end! This choice is determined by:

the scientist's knowledge of completed science;

the scientist's character and disposition.

The factors governing the scientist's choice lead equally to the admission that research is subjective. The first element (the scientist's knowledge) is an element of recognition. This recognition depends on the scientist's field.

Example: A biologist will rather seek recognition in medical science, chemistry or physics than in aerodynamics or theology; an economist will rather seek recognition in social science, psychology or mathematics than in biology or chemistry.

The more disciplines the scientist knows and the greater the area in each discipline he understands, the easier is his choice; at least the more he has to choose from. All the same, every scientist knows only a certain part of science and therefore tends to seek his way in a particular direction. The way he seeks depends on his whole character; one feels intuitively for the right direction, another first tries to reduce the problem to its simplest form, a third follows the path of least resistance, and a fourth finds his pleasure in groping in the confusing complexity.

Whatever his principles of choice, the process is only partly a mental one; it is much more a process of intuition (a word for a concept that is not yet understood!), a subconscious or, perhaps better, a superconsciousness (surscience). In this primordial stage the process is not tied to the laws of logic but does have an infra-logical structure. As soon as a step has been taken intuitively on the path, this step is confronted by the purposeful thinking that the step is testing, either in the already known or in the experiment.

In this way, the scientist constructs a plan for the building he thinks he will put up. In this, he shows many characteristics corresponding to those of the artist. The artist chooses the elements (paints, clay, etc.) with which he designs what will suit him. The scientist chooses the elements that suit him for the construction of the plan. Both work intuitively at this stage. Once the scientist has completed his plan, in other words reached his conclusions, hypotheses, or theories to which his research leads, he is all ready for the job of starting the real building. That means the research must be transferred from the stage of science in creation to the stage of completed science. This happens in the publication.

This discussion shows why the publication cannot be put together along the same lines as the process of research so far described. From the meandering process through the labyrinth of the unknown, with repeated testing, a logical construction must *afterwards* be built, progressing step by step in the research report. The start and finish of the process of research and of publication are identical: in both, we start with the definition of the problem and finish with the conclusions. But the process of research meanders from junction to junction between these two points and in the publication it must run straight (Fig. 16).

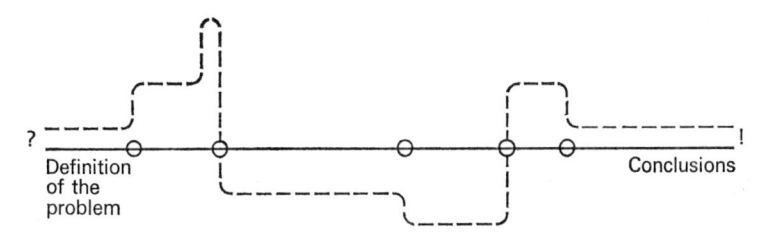

Fig. 16. Process of research and scheme of publication. Broken line—process of research; continuous line—scheme of publication; circles—paragraphs.

This transition from science in creation to recorded science in the publication requires a good deal of effort from many scientists; it requires an entirely different disposition from that of research. In research, there is a phantasy of playing with the elements available; in publication, there is an exact and logical reasoning with a confrontation, with doubt at every moment and, as a consequence, the supply of evidence. Furthermore, in writing the report a choice must continually be made between essential and inessential and one must resist the temptation to include every detail; a straightening of the line thus and a condensation of the material to the essential.

8.7 Expression of numeric data

The data obtained from observations or analyses are generally expressed in numbers and units, in other words quantities. Research is directed towards clarifying the relation between these quantities. To reproduce the structure of these relations, tables or graphs are often used. We will discuss some elementary principles of tables and figures below.

8.7.1 Tables

Definition

A table is an ordered arrangement of numbers or, exceptionally, words whose purpose is to show the relation between the values of these numbers or words. Tables are usually constructed in columns. In constructing tables, this purpose must be kept well in mind. One must therefore continually ask the following questions:

Can the table be arranged more simply for a general impression?
Can the numbers be arranged in another way to demonstrate the relationship more clearly?
Can the numbers or descriptions be simplified?

EXAMPLES:

(*i*) A table with many columns is difficult to read. It may be advisable to split such a table into two or more.

(*ii*) We may tend to arrange data from field trials in a traditional way from north to south or from west to east or chronologically. But we should question whether there are better ways of demonstrating relationships between values.

(*iii*) If the values are rounded off in thousands they can be simplified in the table by taking a unit one thousand times greater (for instance tons instead of kilograms or kilograms instead of grams). Alternatively the quantity measured can be divided by a thousand (thus instead of k one may measure $k/1000$).

(*iv*) One must examine which is clearer: a broad table with few lines or a long table with few columns.

Elements of a table

Tables should be so made that they can be read apart from the text. Each table must therefore have a *caption* carefully describing its content. A caption may be only a title consisting of an incomplete sentence without a verb, or following this title may be further sentences of explanation. It is advisable to number all tables in a publication or report in sequence and to give the word Table as Table 1, Table 2, etc. An exception is for very simple tables consisting of not more than two columns which are actually included in the text. They can be called *enumerations* and have no number.

EXAMPLE:

'In recent years foreign trade by Netherlands Antilles has had the following trend:

	1961	1962
exports	1337	1297

imports   1352   1358
Export of oil and oil products is by far the most important commodity.'

A table is divided vertically into *columns*; the first column generally consisting of text is called the *stub*. This column defines values in the rule or line. Sometimes such a column is placed on the right, the right stub (see under mirroring of tables); sometimes the columns are numbered. Each horizontal line of data is called a *rule* or row; the whole area that can be filled with data is called the *field* and consists of co-ordinates containing items. Above each column is a column *head(ing)*. In it is a description of the quantity in the column and the unit in which the values in the column are expressed unless this quantity or unit is already stated in the caption. It is now recommended that the heading should consist of a symbol written in italic letter for the quantity followed by a solidus and then by the symbol for the unit used or, for example, a multiple of the unit used (e.g. m/kg or $m/1000\text{ kg}$). The more traditional practice is to name the quantity in words (often abbreviated) followed by the symbol or the unit in round brackets (e.g. Yield of hay (tonnes)). A head(ing) is sometimes further divided into subheads. The whole collection of column head(ing) and subheads is called the *boxhead* of the table. The different terms are contained in the model of Fig. 17.

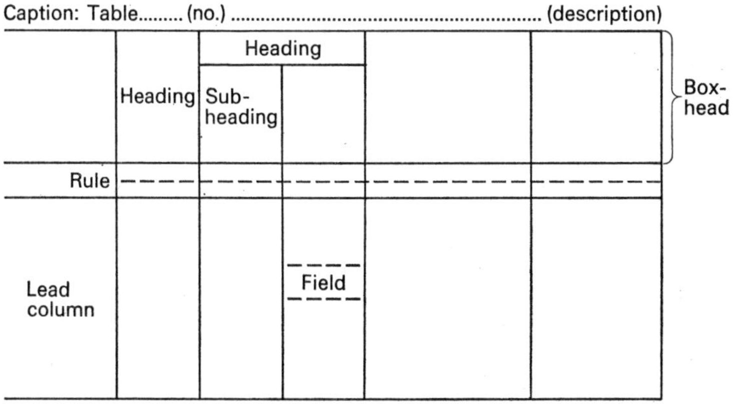

Caption: Table......... (no.) .... (description)

	Heading	Heading			Box-head
		Sub-heading			
Rule	----	----	----	----	----
Lead column			Field		

Fig. 17. Model of a table.

Tables may be distinguished as single entry, double entry or multi-dimensional. A table with *single entry* contains a collection of data analysed for only one characteristic. A table with *double entry* is divided up according to two characteristics. This is the most usual

141

arrangement. Since we work with paper we can only arrange values in two directions: horizontally and vertically. If we wanted to go further we would need a third dimension. We can, however, get around this by repeating one of the subdivisions. In this way we obtain a multi-dimensional table. Such tables are, however, not very easy to use.

Example: We want a table of the constitution of a population group over the years, by sex and age group according to sex. We then have three characteristics and can build up the following scheme for them:

Caption: Table ... Population group ... over the years ... according to sex and age

Lead column	men				women				total			
or stub	age groups				age groups				age groups			
years	A	B	C	D	A	B	C	D	A	B	C	D
column no. 1	2	3	4	5	6	7	8	9	10	11	12	13

If we wish to examine Age Group A, we have to compare Column 1 with Columns 2, 6 and 10. The columns do not adjoin! For Age Group B, C and D, there is the same problem.

Types of table

Tables can be distinguished into those for one's own use and those for others (exemplified by those in a publication or report). Remember that these tables for others must be read and understood by people who have not constructed the tables! The first tables derived from an experiment are tables *of observations.* These are simply collections of data. From these tables of observations, *documentary* tables must be compiled in which the data are arranged in a way that can be surveyed and according to a certain order (for instance from north to south, from large to small, from many to few, from good to bad, chronologically or by soil type).

One sets to work with these tables: calculates averages, processes the numbers mathematically and so forth. Hence originate *working tables.* Only then can one decide which tables to make for others: *demonstration tables.* These tables contain only a few data and must serve to show the reader a certain phenomenon clearly. *Demonstration tables must be framed in the simplest way possible.*

Finally there are tables *for consultation.* These tables are for looking up data, for instance tables of logarithms or of nutritive value of foodstuffs and timetables. For such tables, close attention must be paid to the ease with which data can be looked up.

Construction of tables

The sequence of lines and of columns largely determines the impression that the table gives. In general, the most important data are placed in the first column. But one must also remember that comparison of figures is easiest in adjoining columns. If possible, one must also try to make the values in at least one column run from large to small or small to large. The values must be reduced to the fewest digits possible. Never place more figures after the decimal point than is reasonable.

If one is working with averages, it will often be necessary to indicate the scatter (for instance by adding standard deviation). Numbers above 5 may be rounded upwards and below 5 downwards. After rounding, the totals may not necessarily agree. To avoid bias of averages made from rounded numbers values ending exactly in 5 may be rounded to the nearest even number, for instance 0.345 is rounded down to 0.34 and 0.335 is rounded up to 0.34.

Other conventions in tables are as follows:

1939–1940 means 1939 to 1940 (American 1939 through 1940)

1939/1940 means a year beginning in 1939 and ending in 1940 or another period beginning in 1939 and ending in 1940, for instance the winter of 1939/1940.

. means data lacking

—means data is precisely 0

0(0.0;0.00) value too small to be expressed

* value tentative

Blank: the value cannot exist

$\bar{x}$ (1939 + 1940) means the average of 1939 and 1940

If entities are arranged in classes according to the values of parameters or according to attributes, a rule that is useful is that the number of entities can be divided into 2 to 5 times the logarithm of the number of entities, for instance with 100 entities or observations one can have 2 to 5 times log 100, that is 2 to 5 times 2 which is 4 to 10 classes.

The arrangement in classes must be clearly stated, for instance, if divided into groups with a range of 100, the headings to the columns must not be 1–100, 100–200, 200–300, etc. but 0–<100, 100–<200, 200–<300, etc. As long as English is the main language of the text, use a point for the decimal. Remember that for other languages the comma is used. If there is a column of figures all expressed in one

unit the number of digits after the decimal must always be the same. Not, for instance, 2.25 but 2.25
12.16 12.16
13.4 13.40

In each direction from the decimal place, numbers are divided into groups of three digits separated by a space. Avoid the use of either comma or point as a divider. For instance: 12 397; 1 105.7; 2.397 6.

Tables in the text should preferably not contain more than 12 lines and 6 columns. Larger tables are better placed at the end of the publication as appendices. Try to push information as far upwards as possible. In other words, from the columns into the subheads and from the subheads into the heads (for instance, do not place the £ sign with the number in the field but place it in the heading); likewise take matter from the boxhead into the caption, if, for instance, all units are expressed in grams per cubic metre place them in the caption and not in the boxheads.

In large tables a white line between every five lines of figures improves the readability. Usually the readability of tables can be improved by using vertical lines as little as possible (although this has been done for clarity in the schemes of Figs. 17 and 19), the belief that a vertical line between columns separates them better than white space has proved false. On the contrary, the careful use of only horizontal lines (as a sort of brace) can improve the readability of tables; Fig. 18 is an example.

Table 29. Numbers of actinomycetes and pathogens from soil in irrigated and unirrigated parts (PAW 1057, 1967).

Treatment[1]	On GAT-medium				Pathogens 22 June ($\times 10^3$)	Scabby tubers (%)
	total act ($\times 10^3$)		tyr+ act ($\times 10^3$)			
	22 June	7 July	22 June	7 July		
V_1	960	1030	240	410	14	12.7
	—	1200	220	650	11	7.7
	890	1100	490	460	7	6.0
	920	700	340	340	5	9.6
V_0	890	1440	480	830	20	48.5
	840	2810	490	1340	21	54.5
	940	3110	480	1340	29	57.8
	470	2540	270	1230	4	53.4

1. V_1: treated with 65 mm water in total, V_0: untreated

Fig. 18. Example of a table with horizontal lines.

In tables with two languages, the text can often be *mirrored* to improve the typographic balance of the table (see Fig. 19). If a table includes footnotes, it is advisable to place them directly under the table and not at the foot of the page. For clarity, these footnotes should avoid index numbers but use either signs or letters of the alphabet.

Table ++	Title +++++++++++++++++			
++++++++	+++++	++++++		
		+++	++++	
++++	------	---	-----	oooo
+++++	---	--	---	ooooo
+++	----	---	---	ooo
++++	----	-	--	oooo
	ooooo	ooo	oooo	oooooo
		oooooo		

Table oo   Title ooooooooooooooooo

Fig. 19. Model of a mirrored table. Crosses—Dutch; circles—English; dashes—values.

8.7.2 Graphs

Definition

A graph is a drawing made up of points, lines or surfaces whether or not supplied with figures, letters or words, by means of which quantities can be demonstrated visually. A graph may also be called a graphic representation or a diagram. The purpose of a graph is usually to *illustrate* the relation between the quantities under consideration. Since a graph is a representation, it is abstract: a graph never reproduces reality in the same way as a photograph or a working drawing. Even pictorial statistics, in which quantities are represented by small figures (puppets, animals, boxes, etc.) are abstract. This abstract picture can easily give rise to incorrect conclusions; a graph can be misleading. Tables are more 'honest'. However, graphs are often easier for people to understand.

Generally one does not use graphs where it is necessary to read off values accurately (as in a table); the purpose is the visual representation of a relationship. This representation must therefore be allowed to come out as much as possible and all superfluous elements must be left out. Superfluous elements in a graph can be, for instance, the stating of all values of the units on the axes, the dividing of the graph into squares by horizontal and vertical lines, or explanatory texts in the graph.

Types of graph

There are many types of graph. We will discuss the most common ones.

PICTORIAL GRAPH

A pictorial graph is used only to represent very simple statistics (for instance the number of beds in hospitals over the years). The unit is represented by a stylistic picture (Fig. 20). A number of pictures (horizontally) next to one another makes a sort of column and the length of the columns can be compared with one another. The pictures (symbols) come under copyright and may not be reproduced without permission. In scientific publications, pictorial graphs are rarely if ever used.

Example of a pictorial graph

= £2 000 000

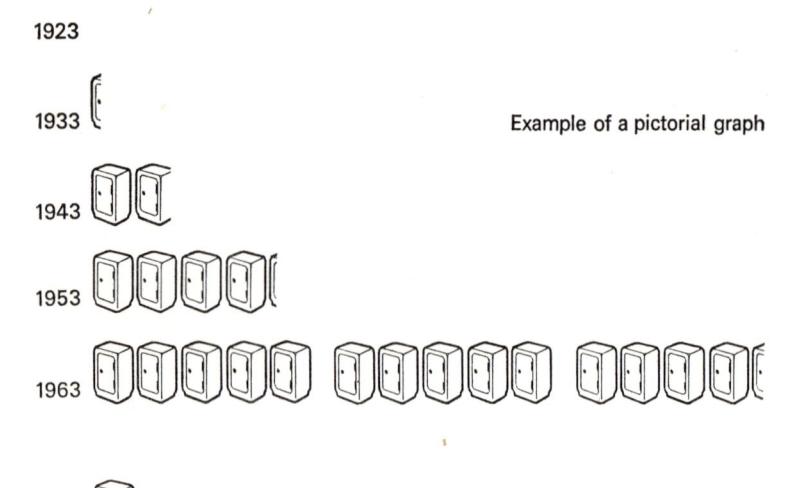

Fig. 20. Example of pictorial graph.

CIRCULAR GRAPH

A segmented circle is a circular graph in which the area of the segments represents the quantities (Fig. 21). The circle itself forms the total. Totals can be compared with circles of different radius;

the quantities by comparison of segments. Visual comparison of segments is not very easy. Circles can therefore only be used for simple and rough comparisons. A second function can be introduced by varying the radius of each segment.

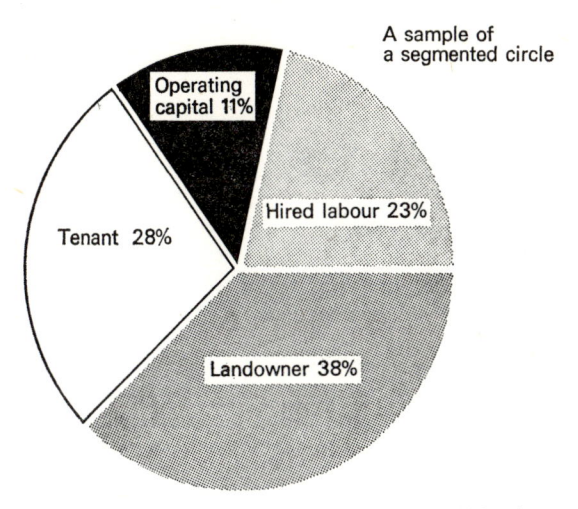

A sample of a segmented circle

Operating capital 11%
Hired labour 23%
Tenant 28%
Landowner 38%

How increased returns on 43 farms in Laguna, Philippines are divided among the four claimants.

Fig. 21. Example of a circular graph.

COLUMN DIAGRAM

In column diagrams the values are arranged as separate vertical columns alongside one another (Fig. 22). According to the distance between the columns we can arrange them, for instance, by time (for instance 1 cm for 1 year; if a year is omitted, 2 cm can be used). For a simple statistic, comparison is easy and clear. Remember, however, that the thickness of the columns influences the impression. The thicker the columns the shorter they look. If one wishes to divide the columns, there are difficulties because the parts cannot easily be compared with parts of other columns, except for the bottom part. The others are at different vertical heights.

HISTOGRAM

If the columns are placed adjacent to one another we have a special type of column diagram, called a histogram (Fig. 23). A histogram can be used, for instance, to place the parts of one column *next to one another*. The columns representing the parts are certainly now

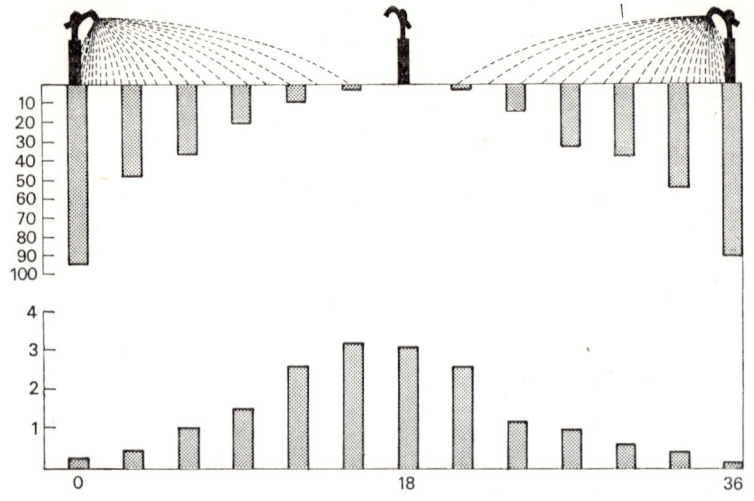

Correlation between the quantity of irrigation water and scab incidence on Sirtema tubers at Ens 1966. At the top, each column represents the total amount of water (mm) during six irrigation occasions which fell between two open and one closed sprinkler nozzles placed 18 m apart. At the bottom each column represents the average percentage surface area of tubers scabbed, when harvested from the different positions between the nozzles (all figures means of five replicates).

Fig. 22. Example of a column diagram.

comparable (on the horizontal axis) but the totals are not. One can also link the top centre of the columns by a line and then leave out the columns themselves; this gives a broken line (line diagram) as in Fig. 24. It is certainly clear but the danger is that one may interpret the quantity between two points as readable from a straight line, which is not necessarily true.

LINE DIAGRAM

If we assume that there is a relation between two quantities and have made some observations, we can present these observations as points in a system of horizontal and vertical lines and connect them by straight lines or by a broken line. We can do this for several relations. In this way we obtain several lines in one graph and can compare them with one another (Fig. 25).

The image presented by a line diagram depends very much on the scales that are used. A great deal can be 'proved' by tendentious choice of scales. Most line diagrams have a horizontal and vertical

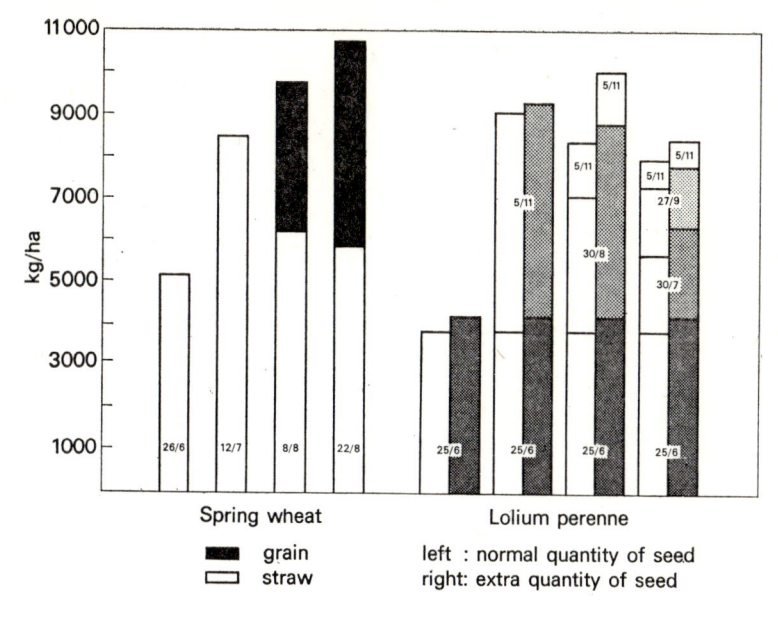

Dry weights of the shoots

Fig. 23. Example of a column diagram (left) and a histogram (right).

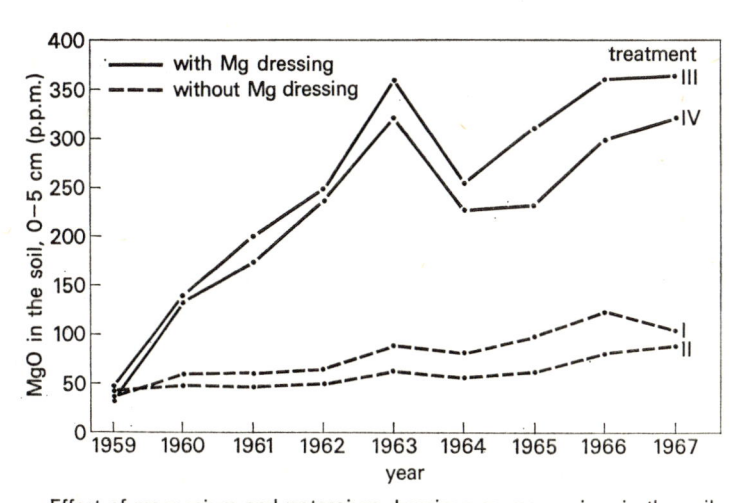

Effect of magnesium and potassium dressings on magnesium in the soil.

Fig. 24. Example of a graph with broken lines.

line meeting at a point. These lines are called the axes or co-ordinates. The horizontal axis is the *x* axis or abscissa; the vertical axis is the *y* axis or ordinate. The units are set along the axes from left to right and from bottom to top. The intersection of the axes

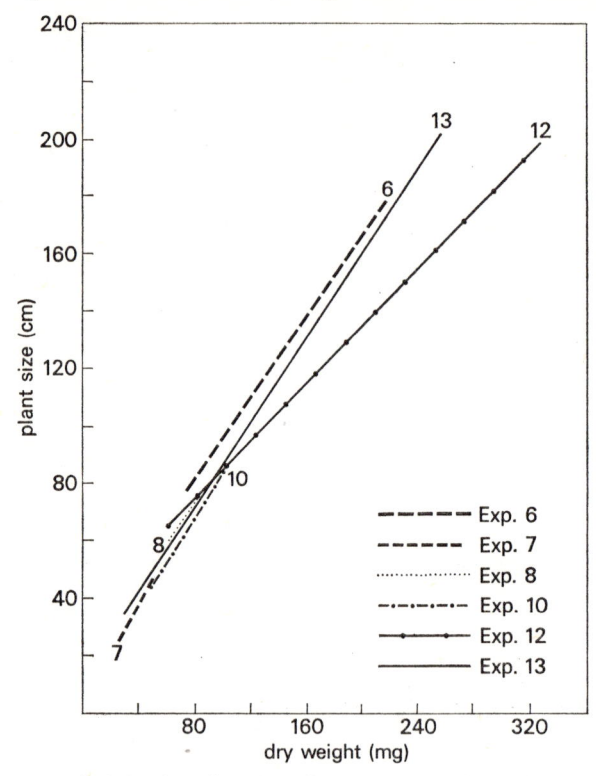

Relation between plant size and dry weight, as found in different experiments.

Fig. 25. Example of a line graph.

is called the *origin*; this is usually *zero*. Sometimes the bottom of a graph is omitted and may, for instance, begin at a value of 120. This may be indicated by a zigzag line in the axis.

The *name or symbol of the quantity* and the symbol of a unit is given near the axis. As far as possible, this should be printed horizontally at the right under the end of the *x* axis and at the top on the left above the *y* axis. The values of the units are indicated by small notches pointing either inwards or outwards on the axes. The way in which units are divided in the axes is called the *scale*. The scale may, for instance, be 4 mm per unit. Sometimes, for instance with an exponential curve, one axis can be arranged on a logarithmic scale so that the curve becomes a straight line (Fig. 26). The values of the quantities can be indicated in different ways, for instance by points, triangles, circles or crosses, all of which are called *points*.

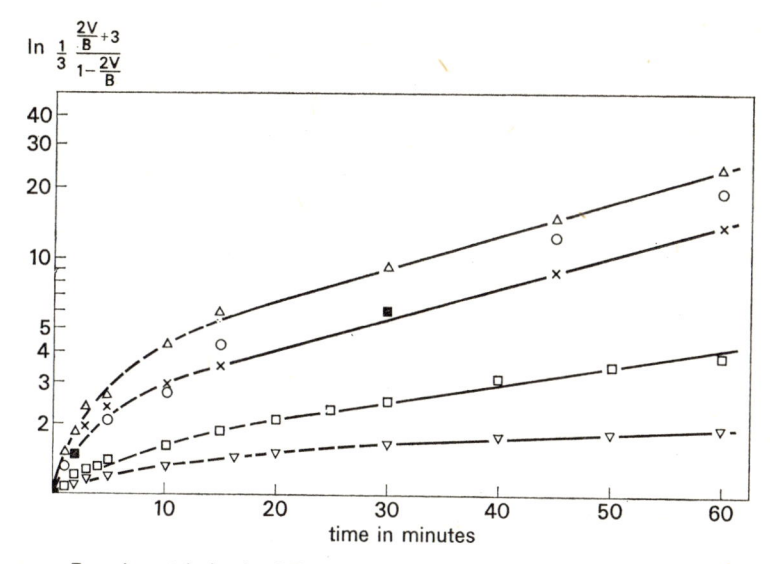

$$\ln \frac{1}{3} \frac{\frac{2V}{B}+3}{1-\frac{2V}{B}}$$

Experimental check of Eq. 16 with five whole milk spray powders.

Fig. 26. Example of a line diagram with a logarithmic y axis.

SCATTER DIAGRAM

It happens sometimes that there are a lot of observations that cannot be represented in any other way than as points in a system of axes. In this way, we obtain a scatter diagram. As such they are difficult to read, and lines are therefore drawn (often by eye) either in a curve or straight through the points (Fig. 27). Here, too, the reader must be very careful. One can easily draw a thick line through points and suggest a relationship that does not really exist.

TRIANGLE DIAGRAM

Sometimes one wishes to compare three components. A solution is the triangle diagram (Fig. 28). For this one considers the total of the three components as 100 and calculates the percentages of each component for each separate observation or analysis. This observation results in a point in the triangle. If this is done for a large number of observations or analyses a collection of points is usually

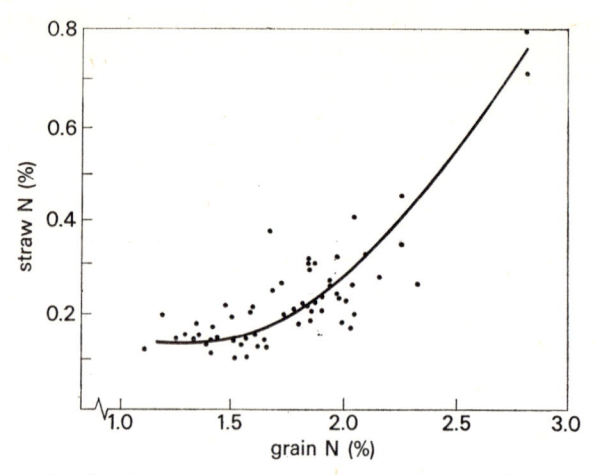

Relation between nitrogen in grain and in straw. Data are of N_0 and N_1 treatments.

Fig. 27. Example of a scatter diagram with a zigzag on the x axis.

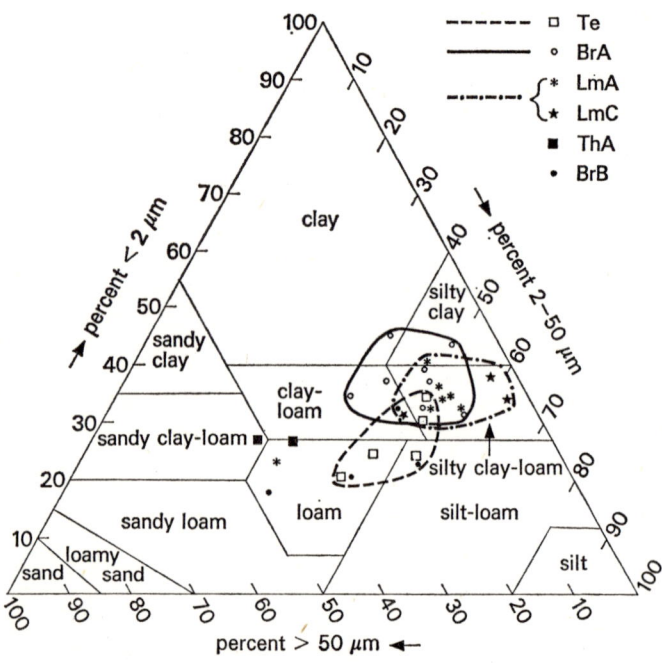

Particle-size distribution of individual samples from the plough layer.

Fig. 28. Example of a triangle diagram.

152

obtained within the triangle. The position of this collection allows conclusions to be drawn.

THREE-DIMENSIONAL GRAPH

Like the table, the graph has the disadvantage that one is working in the flat, thus in two dimensions. It is possible to overcome this to some degree by drawing a third axis at an angle of 135° downwards from the x and y axis. One obtains a representation of three axes at right angles to one another in space and can construct graphs in the form of surfaces. The spatial relation can thereby be suggested. The interpretation of such graphs is often far from simple, and are therefore not used very often (Fig. 29).

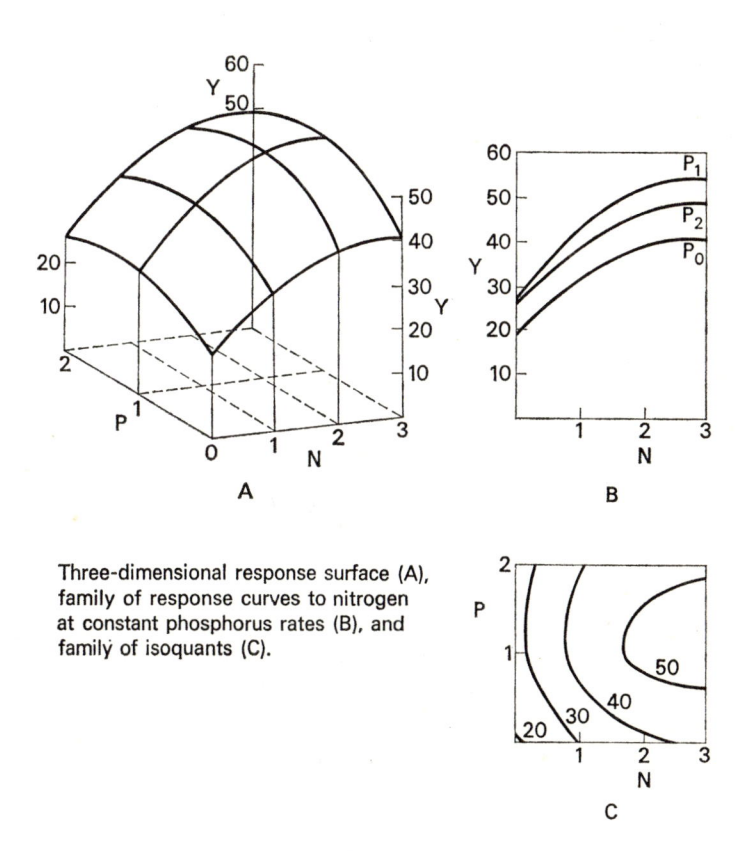

Three-dimensional response surface (A), family of response curves to nitrogen at constant phosphorus rates (B), and family of isoquants (C).

Fig. 29. Example of a three-dimensional graph.

153

NOMOGRAM

In contrast to other graphs, which only give a visual and abstract representation of a relationship, the nomogram is intended for reading of values. The nomogram corresponds therefore to reference tables and many tables of reference can indeed be converted into nomograms (for instance a table of logarithms or a railway time-table). A slide rule is in fact also a (movable) nomogram. A nomogram must of course be very carefully drawn. One can only use it after studying its principles carefully. Usually one has to connect two points on two axes to read off the information required on the third line (Fig. 30).

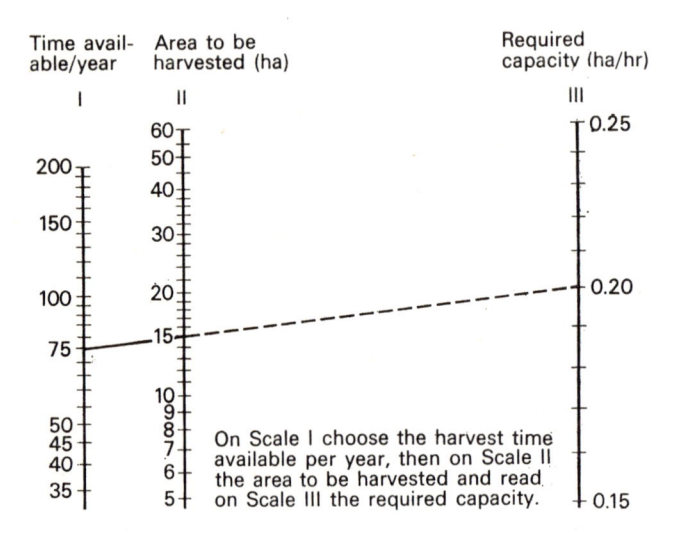

Fig. 30. Example of part of a nomogram.

FLOW-DIAGRAM

A flow-diagram is sometimes used to represent the lines along which a product is processed. The thickness of the flow lines represents the amounts of product and its derivatives. The processing of the product, for instance the raw material, is represented by a square, a sort of lock in the flow (Fig. 31). The representation can be very illustrative but must be kept very general.

NETWORK DIAGRAM

An entirely different sort of graph is the diagram for networks, linear programming and such like, which are used in many discip-

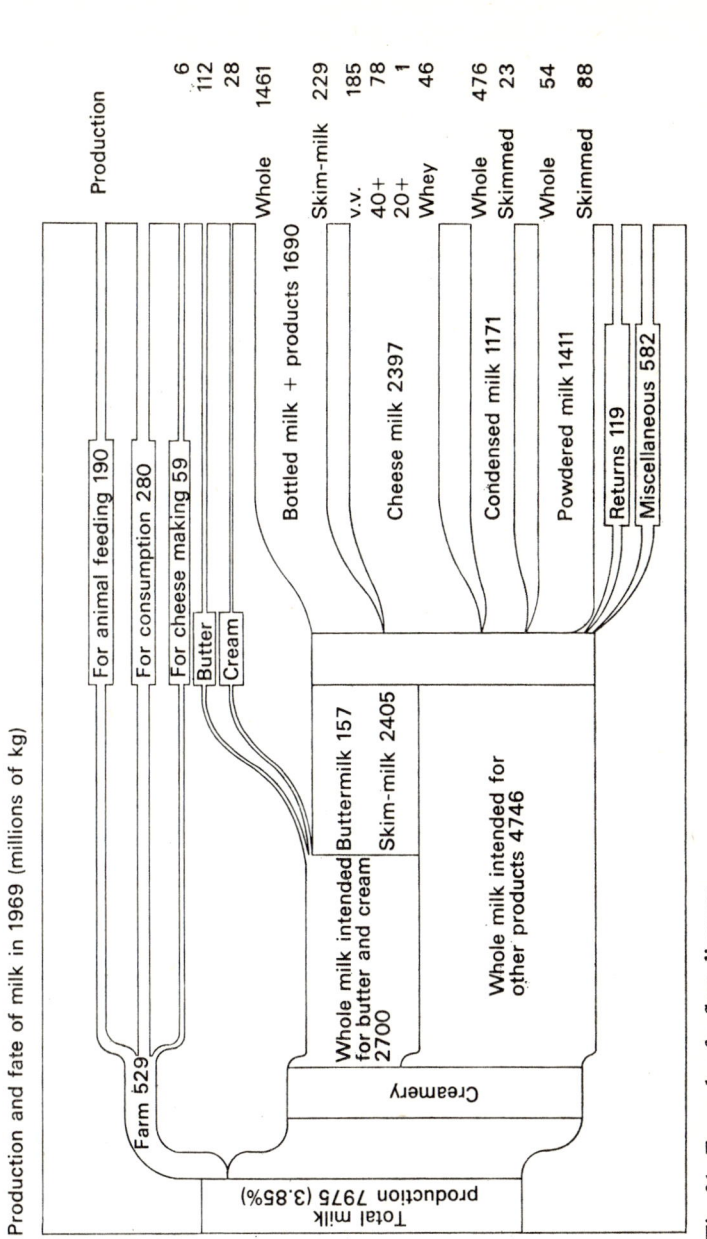

Fig. 31. Example of a flow diagram.

lines. There are links between points by lines or arrows. The points can represent different things: persons, places, chemical materials, genetic properties, etc.; the lines or arrows equally: family relationships, lines of organization, pipelines, railway lines, etc. (Fig. 32). Theoretical principles for these diagrams have been constructed with a distinct nomenclature based on Boolean algebra.

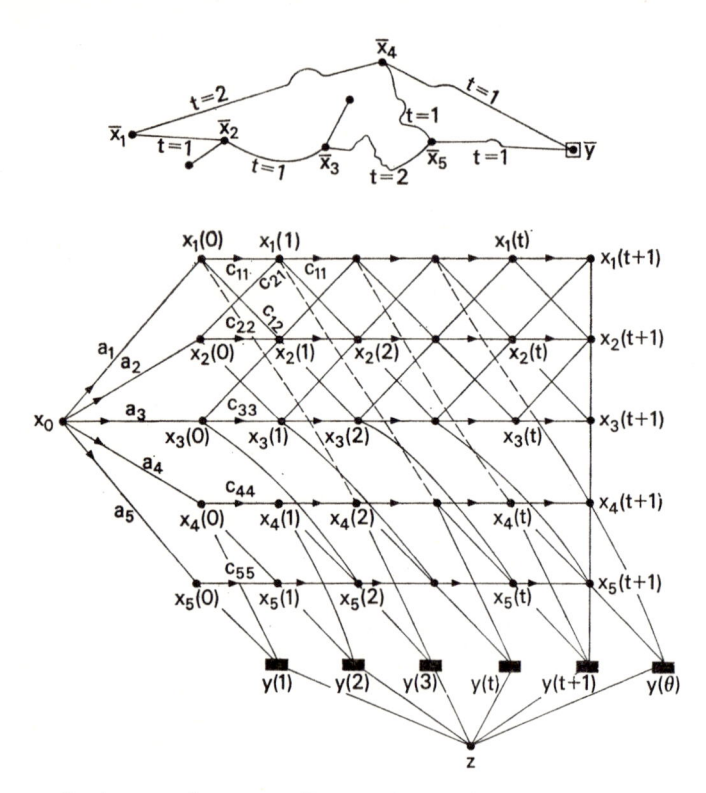

Road map and corresponding transport network in space-time.

Fig. 32. Example of a network diagram.

Preparation of graphs

The drawing of graphs for publication (for preparation of plates or blocks) is skilled work, which should be left to a skilled draftsman. It is advisable to give the draftsman a rough form on graph paper and let him work it out into the form for the printing plate or the block. It is of course necessary to check this drawn diagram very carefully before it is photographed for the plate. It is almost impossible to make corrections in a plate or block!

Further reading

Beardsley, M. C. & Beardsley, E. L. (1965). *Philosophical Thinking: An Introduction.* Harcourt, Brace & World; N.Y.
Berge, C. (1962). *The Theory of Graphs and its Applications.* Methuen; London.
Bochénski, I. M. (1959). *Die zeitgenössischen Denkmethoden.* Francke; Bern.
Bronowski, J. (1961). *Science and Human Values.* Hutchinson; London.
Collingwood, F. J. (1961). *Philosophy of Nature.* Prentice Hall; Englewood Cliffs, N.J.
Harré (1970). *The Principles of Scientific Thinking.* Macmillan; London.
Parker, R. E. (1973). *Introductory Statistics for Biology.* Arnold; London.
Peterson, M. S. (1961). *Scientific Thinking and Scientific Writing.* Chapman & Hall; London.
Popper, K. R. (1972). *Objective Knowledge: An Evolutionary Approach.* Clarendon; Oxford..
Price, D. J. de Solla (1965). *Little Science; Big Science,* 2nd edn. Columbia University Press; N.Y.
Pyke, M. (1961). *The Boundaries of Science.* Harrap; London.
Salmon, S. C. & Hanson, A. A. (1964). *The Principles and Practice of Agricultural Research.* Leonard Hill; London.
United States Council of Biology Editors (1972). *Style Manual for Biological Journals,* 3rd edn. Am. Inst. of Biol. Sci.; Washington, D.C.
Weigel, G. & Madden, A. G. (1961). *Knowledge: Its Values and Limits.* Prentice Hall; Englewood Cliffs, N.J.

References

Crombie, A. C. (ed.) (1963). *Scientific Change.* Heinemann; London.
Janse, A. R. P. (1969). *Sound Absorption at the Soil Surface: A Theoretical Approach with Some Experiments.* Pudoc; Wageningen.
Moles, A. A. (1957). *La Création Scientifique.* René Kister; Geneva.

Sittig, J. & Freudenthal, H. (1951). *De Juiste Maat* (The Right Size). Publisher unknown.

Zouwen, J. van der (1971). *De Probleemstelling als Probleem* (The Definition of the Problem as Problem). Samson, Alphen a/d Rijn; Netherlands.

9 Disseminating the results of research

When dealing with the information chain, it was pointed out that most results of scientific research must find practical application (Section 1.4). The only research that does not require this is research intended for gathering knowledge or for supporting other research (fundamental research). Means must thus be found of disseminating research results.

In large industrial concerns, there is generally a fairly direct and short line of communication between the research department and management. The results of research can be applied—so far as this is commercially justified—along these lines in the concern itself and likewise so can those taken from the literature or from patents. In small industrial concerns and also in the agricultural branch of industry, one does not have one's own research department. In many countries, the government pays all or part of the costs of research or, for instance, through a research council or by itself (for instance in agricultural research institutes under the Ministry of Agriculture). We need not here discuss the reasons for this (Arnon, 1968; Maltha, 1968).

The advantage, which should not be underestimated, is that the results of this research are public property, and the government or the bodies subsidized by the government purposefully try to communicate these results to industry or to the public. The disadvantage is that the lines of communication, for instance between research and industry, are longer so that there is a chance that the connection between results and application is lacking. This problem has been recognized and attempts have been made to bridge this gap in different ways. Below we will confine ourselves to written publications and thus ignore many other communication media, such as lectures, films, radio and television transmissions, exhibitions and advisory meetings and open days at research institutions.

9.1 Lines of communication

There are two lines of communication along which the results of research can pass:
(*i*) direct information from scientist or research institute to the industry;
(*ii*) indirect information to the industry through an intermediary group.

9.1.1 Direct information from research to industry

The scientist's primary job is research and the recording of his results in a report. Since little usually remains of the research itself (e.g. test tubes are emptied, equipment is dismantled, trial fields are ploughed up), the recording must be so done that there is no doubt later about how the research was done and what the results were. In other words, the report must be such that with it alone the research can be repeated and the results recalculated. This means precise and detailed reporting, even if the scientist as author tries to condense matters.

In research one usually builds on what one's predecessors have found. If a scientist makes mistakes in drawing his conclusions or draws a conclusion that is premature, there is a large chance that others will build on what proves to be quicksand. The scientist is aware of this responsibility and tries therefore to formulate his conclusions, carefully avoiding any risky statement. This would put his good name as scientist and that of his institute in jeopardy if he should draw a conclusion that is unwarranted. These two factors make the research report less suitable in general for direct transmission to industry. A further problem is that these reports, through their specialized character, are also too difficult for industry to accept the content as such and to apply it.

It is even questionable whether it is right for the scientist to attempt to write articles for industry or farming himself. In the first place, the scientist must have a disposition towards research and not towards advice. In the second place, his time is expensive, not only because he has a high salary but also because a lot of capital is invested in equipment and instruments for his research. Only exceptionally, therefore, should the scientist write articles for industry himself. This can happen, for instance, if the research institute where he works is closely involved with industry (for instance horticultural experimental stations or a research station for the graphic industry) or if industry requires direct information in an emergency (for instance in a severe plague or if there is a change in regulations abroad). But it is not the normal work of the scientist to write practical advisory articles.

Taking this as a premise, the scientist must agree that such articles are written by someone else. The scientist should never send a research report as such for publication in a trade journal. If information is to be given directly to industry this must be done by one of the institute staff or by a reporter from a trade journal. Of course these writers must ensure that the scientist accepts the article even though he does not bear responsibility for it. In referring to the research, the scientist's name must be mentioned. 'Honour where honour is due' rather than 'a feather in one's own cap'. A good working relationship is thus necessary in which face-to-face discussions are to be recommended.

If someone at the institute has the job of writing and editing this sort of advisory publication, he must have easy access both to the scientist's work and, in the other direction, to the wishes of industry. Further, he should build up relations with the publicity media (such as trade journals) so that he can place the articles he writes. It is good policy to gain a regular column in suitable trade journals. It is advisable also to maintain contacts with trade journals in other ways, for instance by inviting editors of trade journals and reporters to visit the institute, to learn for themselves about one or more aspects of the research the institute is doing. Such liaison must be well prepared, and sufficient written material should be available for the visitors—but do not overdo it. Liaison with the press can also be encouraged by passing on tips. Such tips must be chosen carefully: they must contain an element of news and not all research results are suitable for this. These tips are not intended for publication as such (although it often happens), but to set the reporters or editors of the journal on the track of interesting developments. Another form of contact is the press release intended to be published as such in trade journals and newspapers.

If the clients of a research institution are quite few and are known (for instance manufacturers of certain types of machines), the institute can approach these people or institutions directly through its own publications (for instance standardized notices or small brochures). Sometimes one can go a step further and circulate reports in the form of a house journal.

9.1.2 Indirect information to industry

In most countries, special advisory services have been built up for agriculture and industry. Among their tasks is to ensure that the results of research—so far as they can be used—find application. These advisory services are organized differently in different countries. Sometimes they are close to the research institutions, as in the United States. The advantage is that they can quickly and easily get hold of results from scientists. A disadvantage, however, is the distance from the industry and these services have less easy contact with the firms. Sometimes the services are at the other extreme, heavily dispersed and are close to the firms. But it is then more difficult for advisory staff to keep up to date with research.

It is reasonable to say that it is the scientist's job to pass information about his findings to advisory workers in a suitable form for them. This can be in the form of articles in suitable trade journals. Another way is as a simple bulletin, for instance in stencilled form. One must also take care to supply information required verbally. Information must pass in both directions: from the research institution to the advisory service and so to the firms and in the reverse direction through the advisory service to the scientific institutions (trouble shooting). Exceptionally when the research is particularly directed towards industry or the farmer or market gardener the primary research report, perhaps with some abridgement, can be passed directly to the intermediary group. But usually this is not desirable.

The scientist must then write a *second* report for this group as well as his primary research report. It is desirable that the manuscript be edited by someone who knows the mental attitude of those working in advisory services. The most important elements from the primary report for other workers are:

the reasons for the research;
the problem;
the conclusions.

These three elements must therefore be presented in the publication for related workers. A section must be added after the conclusions stating possible applications. It is advisable also to pay attention to economic aspects of the application.

What is taken from other sections of the research report depends on the subject. In any case, one must simplify and abridge appreciably. It certainly seems important to tell the related worker something about the method used and to give him a brief survey of results.

The conclusions are the most difficult part. To take them over completely may not be right. For the advisory service, they contain too much ballast. But to translate them into a black and white picture is not desirable either. Sometimes it can be preferable to select only some of the conclusions.

The following scheme indicates which elements must occur in the two types of articles. Parts which are optional for advisory services are indicated between brackets.

Primary research report
1. Introduction to the problem
2. Critical review of the literature
3. Definition of the problem
4. Methods of research
5. Observations and actual results
6. Discussion of results
7. Conclusions
8. Summary
9. Bibliography
10. Appendices

Article for extension
1. Introduction directed to industry
2. (Short indication of previous research)
3. Problem studied
4. (Brief indication of methods of research)
5.
   Discussion of results
6.
7. 'Translated' conclusions
8. Indication of practical applications

9.2 Types of advisory publications and their characteristics

9.2.1 General characteristics

Advisory publications for industry are intended to achieve something among the readers. They are not intended only to inform about facts but to influence the behaviour of the reader who must, by reading the publication be influenced to do something, to change his attitude, or something of the sort. One may not expect that this end is reached in all cases; one of the characteristics of advisory publications is that much is lost. It is also often advisable to bring influence to bear through other media, such as demonstrations, films, lectures and discussions. Publications fulfil a complementary role. Advisory publications are part of the mass media. They are relatively cheap. As the number printed is increased, each pamphlet or document becomes cheaper. There is therefore the tendency to print large numbers and to attempt to spread them *en masse*.

On the other hand, it is generally easier to influence the individual's action, the more it is directed to the person. Here lies a conflict. Behaviour can be influenced more easily too by more specific advice. An advisory publication, for instance on poultry keeping, will be less successful than one on how to make nesting boxes. The advisory publication lies between the scientific publication and written propaganda. The advisory publication shares its objectivity with the scientific publication; but it differs in the lesser need for complete evidence and, in that, it must be more black and white than the scientific publication. Like written propaganda or advertising, advisory publications are particularly directed to the influencing of the will. But it must be more objective and therefore trustworthy.

Fig. 33 shows schematically how the research report is intended only to increase the reader's knowledge. This is the only purpose of this sort of publication. The other extreme is the advertisement intended almost entirely for its influence on the reader's will (for

```
Increasing                                    Influencing the
knowledge                                     will of the reader

  \               Research report                 /
   \              Advisory publication           /
    \             Written propaganda            /
     \            Advertisement                /
      \                                       /
```

Fig. 33. Scheme showing how the research report is intended to increase the reader's knowledge.

instance to buy goods of a certain make). Written propaganda is intended to promote a good purpose, at least so in the composer's view. This type of publication tries to win over the reader (in other words to influence his will), but it usually does so by providing information about the cause and thus also increasing knowledge.

Example: a piece of propaganda for a blood transfusion service will give information on the purpose and the way in which the service works as well as influencing people to work as donors for the service. It will therefore only contain positive arguments.

In an advisory publication, both pro's and con's must be given. Experience shows that if this is not done, the reader doubts the objectivity of the advice (he considers it as propaganda). On the other hand, the pro's and con's must not be left in balance; this will *certainly* not convince the reader nor cause him to act.

Searching for the right form of the message makes the writing of advisory publications no easy matter. Each word must be weighed up and considered. The choice of words in an advisory publication must, of course, be within the linguistic knowledge of the intended readers. One can usually assume that technical terms will be understood. Strange words must be used in moderation and with care. They cannot always be avoided; sometimes they must be explained. The style must be brief and direct.

Opinions differ as to whether advisory publications should begin with an introduction to the subject or whether one should walk straight in at the front door. The leading of the reader to the subject has the advantage that it takes the reader from where he is and so increases trust. Getting straight down to the matter leads to more directness and brevity. Either way it is advisable to write the instruction or message in points. There is, however, a difficulty in the choice of forms of instruction. The following examples clarify what is meant:

Shut off the water when you go on holiday (imperative);
We shut off the water when we go on holiday (first person plural);
You must shut off the water when you go on holiday (second person);
People shut off the water when they go on holiday (third person plural);
One shuts off the water when one goes on holiday (third person singular);
The water must be shut off when the occupier is on holiday (passive);

None of these forms is satisfactory in all cases. It is often a question of feeling. Americans prefer the personal approach but it should not be taken to extreme. Take care!

9.2.2 Some types of advisory publication

We will briefly discuss some types of advisory publication.

Article in the trade journal

An advisory article in a trade journal should not be longer than 800 words. It must be well divided up with headings and end (or begin) with clear conclusions. Photographs should be rich in contrast. For particular subjects, it must be published at the right time of year and must therefore be sent in to the editor's office in good time. The title must be businesslike and cover the content; a title playing with words or framed as a question is less acceptable (Section 8.2.5). Use few tables and graphs; indicate how conclusions are reached but do not go back to the very historical beginning.

Sometimes one can obtain a regular column in the journal for short reports. Various studies on reading habits have shown that such features have a wide readership.

Article for the newspaper

Local or regional papers are often more widely read than trade journals. They are thus a good medium. Articles for such a paper must be shorter still than for a trade journal and should contain *news*. News is difficult to define. Journalists frequently quote the well-known example: if a dog bites a man it's no news but, if a man bites a dog, that certainly is.

Pamphlet or bulletin

With selective distribution, the influence of a pamphlet or bulletin can be large. It must reach the reader sooner than an article: the reader must have time to consider the advice before deciding. The preparation of pamphlets takes much time; the text must be written and rewritten again and again; each word must be weighed up. Pamphlets or bulletins can be directed to a whole country or to a region. Regional pamphlets are easier to direct in a particular area than national pamphlets. The local colour increases their acceptability. It is advisable to distribute selectively and not house-by-house. Pamphlets can be distributed by post or can be handed out, for instance, after a lecture. Care needs to be taken in how they are handed out.

Advisory letter

An advisory letter is a sort of circular or collective letter. It should have a personal style and approach the normal letter as closely as possible: with heading and signature. The advisory letter can only be used locally and addressed to acquaintances or members of a trade or farmers' society. But it is an excellent means of advice. The 'you' style must certainly be used.

Monitoring or warning card

The warning card can get straight to the point and can be ready printed as a form. It is a matter of building up trust before such cards will be accepted as an instruction. Preparation can be very simple but one must take care that such cards stand out among other post. The monitoring card is of course only possible in certain circumstances, for instance in control of air pollution or diseases.

Further reading

Aubrac, R. (1973). International co-operation in the field of agricultural documentation. In: *Progress and Prospects in Agricultural Librarianship and Documentation:* (Papers presented at the Regional European Symposium of the International Association of Agricultural Librarians and Documentalists 14-18 May 1973, Wageningen, the Netherlands). Wageningen, Centre for Agricultural Publishing and Documentation (Pudoc); pp. 21–30.

Debons, A. (1974). *Information Science: Search for Identity* (An overview of information—communication science by Walter Store.) Marcel Dekker, N.Y.; pp. 285–97.

Gray, J. & Perry, B. (1975). *Scientific Information.* Oxford University Press; London.

Herner, S. (1974). Trends in library and information sciences. *Science Tech. News,* **28** (1–3) pp. 31–33, 45.

Meadows, A. J. (1974). *Communication in Science.* Butterworths; London.

Pearson, A. W. (1973). Fundamental problems of information transfer. *Aslib Proc.,* **25** (11) pp. 415–21.

Phadnis, S. P. & Sital, R. (1974). World agricultural information and availability to Indian agricultural scientists. *Ann. Libri Sci. Documen.* **21** (1–2) 32–54.

Rothwell, R. (1975). Patterns of information flow during the innovation process. *Aslib Proc.,* **27** (5) pp. 217–26.

Scarf, D. (1975). The future pattern of information services for industry and commerce. *Aslib Proc.,* **27** (3) pp. 80–89.

Waldron, H. J. & Long, F. R. (eds) (1973). Proceedings of the 36th annual meeting of the American Society for Information Science, Los Angeles, Oct. 21–25; Vol. 10. A.S.I.S. & Westport, Conn. Greenwood Press; Washington, D.C.

References

Arnon, I. (1969). *Organization and Administration of Agricultural Research.* Elsevier; Amsterdam.

Maltha, D. J. (1968). *Living for Life: the Interplay between Agriculture and Science.* Pudoc; Wageningen, Netherlands.

Index

The subject index is restricted to terms related to the subjects 'literature search' and 'writing of reports'.
More important reference numbers are given in italics; + means 'and following page(s)'.

Abbreviation, 75, 80
Abscissa, *150*
Abstract (-ing, -or), see also indicative –; informative –; overlap, 5, 7, 8, 12, 13, 20, 24, 29, 30, 45, 47, 50, 51, 60, 61, 69, *77+*, 78, 85, 88, *89+*, 121, 122, 124
Abstract bulletin, see abstract journal
Abstract journal, 2, 8, 12, 15, 20, 28, 33, 38, *42+*, 49, *50*, 60, 61, 69, 77, 78, 103
Academic library, see also library, 1, 2, 9
Acknowledgements, *121*
Acquisition (in library), 9, 16
Acronym, 102
Advertising (advertisement), 164
Advisory article, see advisory publication
Advisory letter, *166*
Advisory publication, 161, *163+*
Advisory service, *161+*
Alerting service, see also current awareness; profile, 53
Alphabetic (catalogue), see also catalogue, 2, 10, 11, 13, 61, *101–3*
Annotation, 21, 30, 47, *77*, *80*, 122
Annual index, 60
Annual report, 2, 19, 28, 29, *49*, 61, 68
Appendix, 30, 104, 121, 122, 124, 126, 144, 163
Archive, *9*
Archive journal, see also journal, *32*
Article, see also key-, primary-, 7, 26, 29, 30, 31, *34–6*, 46

Author, see also corporate –, 26, 27, 48, 51, 58, 61, 102, *121–5*, 160
Author's abstract, see author's summary
Author catalogue, see alphabetic catalogue
Author index, *76+*
Author's summary, see also abstract, 30, 77, 121, 122, **126**, **163**
Axis, see also co-ordinate, *150*, 151, 153

Bibliographic periodical, see bibliography
Bibliographic source, see bibliography
Bibliographic strip, *30*
Bibliographical description, 9, 12, *16–18*, *74+*, *101*, 121
Bibliography, see also national –; secondary literature, 2, 3, 11, 12, 15, 21, *24*, 26, *27+*, 36, 37, 39, 41, *42+*, 48, 49, 58–61, 68, 69, *100+*, 121, 122, 124–6, 163
Bibliography of bibliographies, 28, 48, 58
Binding (report), *105*
Biography (of research), *61*, 69, 123
Boxhead, see also head; subhead, *141*, 144
British Library, 17
British Library Lending Division, 13, 17, 35, 36
Brochure, see pamphlet
Browsing, 19, 69
BTG, 67
Bulletin, see pamphlet; information bulletin

169

Caption, see also head; boxhead; subhead, 99, 100, 104, *140*, 141, 144
Card, see also magnetic –; marker –: Peek-a-Boo –; punched –; warning –, 77, 85, 122
Card file, see also marker card; title card, 2, 9, 12, 13, 23, 28, *49*, 59–61, 66, 69, 72 +, 74 +, 85
Card system, see card file
Category(-ric), 43, 46, 67, 69, 79, 123
Catalogue (cataloguing), see also alphabetic –; central –; chronological –; commercial –; dictionary –; encyclopaedic –; geographical –; international –; joint –; keyword –; national –; special –; subject –; systematic –; union –, 9, *10* +, 12, 13, 16, 19, 28, 33, 60, 61
Catchword, 10
Central catalogue, see also catalogue, 10, 13, 17, 18
Chapter, 86, 90, 103, 104
Chronological catalogue, see also catalogue, 11
Chronological principle (of report), *86*
Circular, see also pamphlet, *166*
Circular graph, see also graph, *146*
Citation, 58, 100, 101, 119
Class (in table), *143*
Classification, see also Universal Decimal Classification, 2, 11, 12, *61* +, 66, 136
Clearing-house, 4, 36
Closing off (report), see also writing of the report, *88* +
Cloze procedure, *99*
Column, *140*, *141*, 143, 144, 146, 161, 166
Column diagram, see also diagram, graph, *147*
Commercial catalogue, see also catalogue, 11
Communication, 159, *160* +
Computer, 4, 10, 12, 13, 21, 23, 28, 31, 40, 42, 47, 52, 53, 57, 135
Conclusion, 71, 80, 86, *88*, 89, 116, 117, 119, 120, *121*, 122–6, 139, 162, 163, 165
Congress reports, see proceedings
Convention (in tables), see also standard, *143*
Co-operate author, *102*
Co-ordinate, see also axis, *150*
Cover, 29, 30,' 75, 104, *105*
Critical (report), *71* +
Cross-reference, 36, 40, 67, 87
Current awareness, see also alerting service, profile, 21, *25*, 42, *49* +
Current Contents, see also current titles, 20, 68
Current research, 14, 44, 68, 116
Current titles, see also Current Contents, *51*

Data-bank, 4, 23, 28, 31
Data-cell, see data-bank
Decimal, *143*
Declassified report, see also report, 34
Definition (of problem, etc.), 116, 119, 123, 126, 139, *140*, 163
Demonstration-table, see also table, *142*
Depository, see also research library, 18
Description (of the experiment), 124
Descriptive (report), *71* +
Descriptor, 52, 62, *66* +, 78, 84, 85, 122, 124
Diagram, see also column –; flow –; graph; network –; scatter –; triangle –, 145
Diary, 119, 125
Dictionary, 91, 92
Dictionary catalogue, see also catalogue, *11*
Didactic principle (of report), *86*
Disc memory, 28
Discussion, 117, 119–21, 124, 163
Dissemination (results of research). 5, *159* +
Dissertation, see also thesis, 36, 37, 49
Documentalist, 2, 12, 35, 102, 121, 136

Documentary analysis, *9+*
Documentary table, see also table, *142*
Documentation, 2, 12, 42, 48, 136
Documentation centre, see documentation service
Documentation service, 2, 7, 8, 11–13, 19, *20+*, 122
Documentation system, 3, 9, 12–14, 24, 28, 60, 61, 66, 69, 123
Double entry (table), see also table, *141*
Drawing, 100, 145
DT, 67
Duplication (of abstracting), *47*

Editor, see also editorial board, 27, 30, 32, 33, 35, 38, 40, 42, 53, 58, 122, 161, 165
Editorial, *30*
Editorial board, see also editor, 26, 29, 30, 32, 35, 41
Encyclopaedia, 14, 28, *40+*, 49, 57, 60, 69
Encyclopaedic catalogue, see also catalogue, *11*
End-matter (of report), *121*
End-note, see also note, *100*
Enumeration, see also table, *140*
Epilogue, *121*
Et alii, 102
Exchange (of publications), 17, 34, 36
Explanation (paragraph), *89*, 104
Extended (title), *123*

Field (in table), see also table, *141*
Figurative speech, *97+*
Figure, 30, 99, 100, 120, 122, 125
Flow-diagram, see also diagram, *154*
Fold-out, *122*
Footnote, see also note, *100*, 123, 145
Free access, *19*
Front cover, see cover

Gate-keeper, *53*
Geographical catalogue, see also catalogue, *11*

Geography of research, 49, *61*, 69
Governmental library, see also library, 17, *20*
Graph, see also circular –; diagram; histogram; nomogram; ordinate; origin; pictorial –; point; three dimensional –; unit, 58, 99, 100, *145+*, 165
Guide (to the literature), 28, *41+*, 49, *54+*, 57, 69, 122
Guideline, see standard

Handbook, see also manual; reference –; 28, 29, 39, 40
Harvard system, *101*
Head(ing), see also boxhead; sub–; running –, 30, 67, 73, 85, 90, 101, 104, 121, 126, *141*, 143, 144, 165
Hierarchy, *62*, 66, 90
Histogram, see also graph, *147*
Homonym, 67
House journal, 32, 161
House rules, 66
Human interest factor, *98*
Hypothesis, 89, 115, *116*, 117, 120, 125

Illustration, 35, *99+*
Impulsive method (of writing), see also writing the report, *83*
Inaccessible language, 27
Index(ing) (word), see also subject –, 4, 5, 11, 21, 28, 34, *42+*, 46, 47, 49, 51, 57, 60, 61, 66, 68–70, 78, 116, 122
Indicative abstract, see also abstract, 30, *77+*, *80+*, 122
Information analysis centre, see information service
Information bulletin, *50*
Information chain, *1+*, *7+*, 159
Information officer, 3, 5, 53, 116
Information service, 6, 7, 19, *23+*, 53
Information specialist, see information officer
Informative abstract, see also abstract, 30, *77+*, *80+*, 122
Insert (in report), *105*

171

Instruction (text), *165*
International catalogue, see also catalogue, 33
International Organization for Standardizing, 105
International Standard Book Number, 76, 103
Introduction(-ory) (paragraph), see also paragraph, 84, *117*+, 119, 121, 163, 165
Inverted pyramid method (of writing), see also writing the report, *84*
Invisible college, 35
ISBN, see International Standard Book Number
ISO, see International Organization for Standardization
Issue, 29, 34, 68, 121

Joint catalogue, see also catalogue, *10*
Journal, see also house –; periodical; reading –; scientific –; trade –, 1, 2, 10, 14–16, 20, 26, 27, 34, 36, 38, 46, 49, 58, 68, 69, 70, 121, 122, 161

Key article, see also article, *60*+, 69
Keyword, see also -in-context; -out-of-context, 11, 12, 21, 42, 45, 51, 122
Keyword catalogue, see also catalogue, 11
Keyword-in-context, *51*+, 68, 69
Keyword-out-of-context, *52*, 70
KWIC, see keyword-in-context
KWOC, see keyword-out-of-context

Language, see also inaccessible language, *79*+, 83, *91*+, *94*+
Leadcolumn, see stub
Learned journal, see journal
Legal depository, 17
Lending, see loan
Letter of transmittal, 90, *104*
Letter to the editor, *30*, *50*
Library (librarian), see academic –; governmental –; national –; public –; research –; special –; university –
Library of Congress, 17
Line (in table), 140, 141, 143
Line diagram, see also diagram; graph, *148*
List of accessions, *50*
List of content, 30, *73*
Literature list, see bibliography
Literature search, see also geography of research; plan of search; retrieval plan; retrospective information, *8*+, *13*+, 35, 40, 48, *57*+, 66, 116, 117
Literature survey, see survey
Loan, 16
Logarithmic scale, see also scale, *151*
Logic(al) (principle), 83–5, *86*, *98*, 104, 115, 116, 119–21, 124, 125, 135, 136, 138, 139

Magnetic cards, see also card, 12
Magnetic tape, see tape
Manual, see also handbook; textbook; reference book, 39–41
Manuscript, 16, 27, 29, 50, 58, 92, 162
Map, 122
Margin, *104*
Marker card, see also card; card file, 73, 85
Mass media, 161, 163
Method (of research), 120, 124, 162, 163
Microfiche, 13, 16, 17, 36
Microfilm, 16, 37
Mirroring (of table), *141*
Model, see scheme
Monitoring card, see warning card
Monograph, 2, 7, 8, 28, 36, *39*+, 47, 49, 57, 59, 60, 69
Multi-dimensional table, see also table, *141*

National bibliography, see also bibliography, 17
National catalogue, see also catalogue, 10

National library, see also library, 17
Network diagram, see also diagram; graph, *154+*
Newspaper, 161, *166*
Nomenclature, 95, 120, 125
Nomogram, see also graph, *154*
Note, see also end –; foot –; preliminary –, *104*
NTC, 67
Number(ing), see also page number, *90*, 100, *101*, *103*, *104*, *140*, 141
Numeric data, see also table, *139+*

Ordinate, see also graph, *150*
Origin (of graph), see also graph, *150*
Output, 52
Overlap (of abstracts), *46*

Page charge, *31*
Page number, 30, 124
Pamphlet, 161, 162, *166*
Paperback, 8, 39
Paragraph, see also explanation; introduction, 85
Patent, 11, 16, 19, 27, 49, 159
Peek-a-Boo card, see also card, 23
Periodical, see also journal, 11, *29*
Photocopy, 13, 17, 34, 36, 78
Photograph, 99, 100, 122, 145, 165
Pictorial graph, see also graph, *146*
Pieces method (of writing), see also writing the report, *84*
Plan of search, see also literature search, *59+*, 69, 89
Point (in graph), *151*
Preface, *104*
Preliminary note, 7
Press, 161
Press release, *161*
Primary (article, literature, publications), 7, 13, *26+*, 28, 33, 36, 38, *49*, 50, 69
Proceedings, 2, 7, 8, 10–12, 14, 28, 29, *37+*, 46, 49, 59, 60, 69, 76
Profile, see also alerting service; current awareness, *52*
Progress report, see also report, 28, *48+*, 116

Project description, 116, 119
Proof correction, *92*
Propaganda, 164
Pro's and con's, *88*, *164*
Psychological principle (of report), *86*
Public library, see also library, 17, *19+*
Publicity media, see mass media
Publisher, 26, 27, 32, 40, 103
Punched card, see also card, 12, 23, 77
Punctuation, *91*, 92

Quick information, *24*
Quotation (marks), *78*, 85

Readable (readability), see also style; reading ease score, 33, 79, 90, 92, *94+*, 144
Reader, 85, 88, 92, 99, 100, 101, 115, 117, 123, 163, 164–6
Reading ease score, see also readable, *98*
Reading journal, see also journal, *32*
Reading room, 41
Referee, 29
Reference, 27, 42, 57, 66, *100+*, 103, 122
Reference book, see also hand –; manual; text –, 1, 7, 8; 14, 20, 27–9, 39, 47, 69
Reference work, *39+*, 41, 49, 57, 60, 66, 69
Report (technical –; research –; literature –), see also critical –; declassified –; descriptive –; end-matter; progress –, 2, 6–10, 16, 20, 26, 27, 28, 31, 32, *33+*, 35, 36, 38, 44, 49, 51, 53, *71*, 86, 115, 131, 139, 160–4
Reporter, 161
Reprint, 1, 11, 19, 34, 35, 49, 123
Reproduction, 35, 36, 99
Research library, see also library; depository, *17+*
Research report, see report
Results (of research), 120
Resumé, see abstract

Retrieval, see also snowball-system, 12, 18, 24, 42, 44, 47, 57+, 60, 66, 123
Retrieval plan, see also literature search, 73
Retrospective information (search(ing)), see also literature search, 21, 32, 41, 68
Review, see also state-of-the-art; survey, 7, 8, 13, 21, *24*+, 27–30, 32, 33, 39, 46, 47, *48*, 49, 59, 60, 119, 124, 163
Rounding, *143*
Row, see rule
Rule, (table), see also table, 141
Running head, see also head, *30*

Scale, see also logarithimic –, *150*
Scatter diagram, see also diagram, *151*
Scheme, *1*, *3*, *4*, *6*, *14*, 115+, *118*, *124*, *126*, *128*, *131*, *132*, *137*, *139*, *141*, *145*, *164*
Science Citation Index, 45, 59, 61
Scientific journal, see also journal, 8, *29*+, 34
Scientific literature, 7
Scope (of research project), 116, 119, 126
Scope note, *126*+
SDI, see Selective Dissemination of Information
Search formulator, *53*
Secondary (publication; source; literature information), see also bibliography, 7, 8, 13, 14, 21, *28*, 29, 36, 38, 40, 49, 57
Section, see also subsection, 85, 86, 90, 103, 104, 119, 120, 124
see also . . . , 67
Selection, *88*
Selective Dissemination of Information, 4, 21, 42, *52*+
Sentence (length), see also style, *94*+, 98, 99
Series, 29, 48, 49, 51, 76
Short communication, *50*
Single entry (table), see also table, *141*
Size (of report), 85

Snowball system, see also retrieval, 26, 45, *58*, 68, 69
Special catalogue, see also catalogue, 38
Special library, see also library, 1, 2, 9, 17, *18*+, 23
Spelling, *91*, *95*
Standard(ization), see also convention, 19, *75*, *95*, *96*, *103*, *105*+, 120, 125, 161
Standard deviation, 143
State-of-the-art (review; survey), see also reviews, 21, 36, 40, 89
Stub, *141*
Style, see also readable; sentence, *79*+, 83, 91, *97*, 98, 104, 122, 165, 166
Subhead(ing), see also head(ing), *90*, 141, 144
Subject catalogue, see also catalogue, 11, 57, 58, 60
Subject heading, 45, 46
Subject index, see also index, 40, 69
Subsection, see also section, 90, 104, 120
summary, see author's summary, abstract
Survey, see also review, 6, *24*+, 26–8, *48*+, 59, 86, 91, 99, 100, 128, 162
Syllable, 98, 99
Synonym, 67, 91
Synopsis, see abstract
Systematic catalogue, see also catalogue, 11, 19, 38
Systematic method (of writing), see also writing the report, *84*+, *111*+, 119

Table, see also demonstration –; documentary –; double entry –; enumeration; field; mirroring; multi-dimensional –; rule; single entry; unit, working –, 30, 40, *99*+, 100, 120, 122, 125, *140*+, 145, 153, 154, 165
Table—
for consultation, *142*+
of contents, *104*
of observations, *142*

174

Tape, 12, 13, 16, 21, *22+*, 28, 42, 44
Technical literature (journal), 7, 8
Technical report, see report
Terms, choice of, *95*
Tertiary publication, 8, 13
Text, 90
Textbook, see also manual; reference –, 20, 29, 39, 49, 57, 60, 69, 88, 120
Theoretical principle (of report), 86
Theory, 86, 116, 138
Thesaurus, *42*, 67, 122
Thesis, see also dissertation, 14, 20, 29, *33+*, 46, 47, 49
Three-dimensional graph, see also graph, *153*
Time-lag, 58, 61
Title, see also extended, 30, 44, 45, 60, 75, 116, 117, 121, 122, *123*, 124, *165*
Title card, see also card file, *72*
Title-page, 39, 75, *104*, 123
Trade-journal, see also journal, 8, *33+*, *161+*, 165, 166
Translation, 2
Transparency, 122
Treatise, *39+*
Triangle diagram, see also diagram, *151*

Trouble shooting, 162
Typography, 33, 90, 145

UDC, see Universal Decimal Classification
UF, 67
Underlining, 90, 104
Union catalogue, see also catalogue, 10, 18
Unit (in table or graph), *140*, 141, 144, 146, 150
Universal Decimal Classification, *62+*, 124
University library, see also library, *17+*, 36

Volume, 29

Warning card, *166*
Words, choice of, see also index –, key –, *95*, 165, 166
Working table, see also table, 142
Writing-the-report (survey), see also impulsive method; inverted pyramid method; pieces method; systematic method; closing off, 59, 74, *83+*, 105, 119, 139

Yearbook, 41

175